Legends
of the Leaf

ALSO BY JANE PERRONE

The Allotment Keeper's Handbook

JANE PERRONE

Legends of the Leaf

UNEARTHING THE SECRETS
TO HELP YOUR PLANTS THRIVE

Illustrations by Helen Entwisle

unbound

First published in 2023

Unbound
Level 1, Devonshire House,
One Mayfair Place, London W1J 8AJ

www.unbound.com

© Jane Perrone, 2023
Illustrations © Helen Entwisle

Text design by Patty Rennie

A CIP record for this book is available
from the British Library

ISBN 978-1-80018-200-4 (hardback)
ISBN 978-1-80018-201-1 (ebook)

Printed in Italy by L.E.G.O. SpA

1   3   5   7   9   8   6   4   2

MIX
Paper | Supporting
responsible forestry
FSC
www.fsc.org   FSC® C023419

For all the *On The Ledge* listeners who believed in
Legends of the Leaf from the very beginning

Contents

Introduction

'When did you first get into plants?' visitors to my home often ask, usually while disentangling a wayward vine from their hair.

The truth is this: I can't remember when plants weren't a source of curiosity and satisfaction in my life. I call it wearing my 'plant glasses': no wispy weed growing in the pavement, no climber romping over a fence, no flower pressed against the glass of a greenhouse is too insignificant to escape my glance.

But my heart really lies with houseplants. Searching for the roots of my obsession takes me back to primary school. In the library, the shelves were draped in yellowing spider plants. My friend Ruth and I must have shown some kind of flair for horticulture, as we were let out of lessons to water these plants back to life. It may have stunted my understanding of arithmetic, but it did set me up for a lifetime of love for gardening. In turn, the spider plants responded to our care by producing many babies at the end of long stalks I later learned to call inflorescences. I started to build my own plant collection, using my pocket money to buy fat-bodied cacti whose sudden spectacular flowers made me gasp.

Forty years on, and my love affair has not abated, but it feels as if everyone else has joined the party. The distinctive leaves of the Swiss cheese plant are all over Instagram, and snake plants are on sale everywhere from Urban Outfitters to Tesco. There are many books on houseplants, covering everything from propagation to styling. And yet most of them remain silent on the matter of where houseplants actually come from, how they have evolved to thrive in those landscapes, and how they have found their way into our homes. Filling in these rich backstories links our specimens to history, culture, botany and horticulture. More than that, it deepens our understanding of their needs. We cannot get houseplants into focus without this context.

Just what makes a houseplant 'iconic'? I selected plants that would be instantly recognisable to most people, whether houseplant obsessed or not: species found growing on windowsills in almost every country around the world, and repeatedly referenced in art and culture, from paintings to films. This is by no means a definitive list. I may have excluded your favourite species – apologies – but I hope this book will inspire you to dig deeper into your chosen plant's stories too.

Many times in the course of this book you will find the words 'we don't know'. While botany has uncovered some of the mysterious workings of leaves and flowers, houseplant species have often been overlooked by scientists: most research has focused on the species that provide us with food and medicine. This is unfortunate, because we are only now discovering some of the potential practical applications of these plants: from decontaminating polluted soil to helping us understand how to collect water in arid landscapes. One of my greatest challenges was delving beneath and beyond the narrative of colonial botany – the white men who fanned out across the globe in the eighteenth and nineteenth centuries to capture, categorise and exploit plants and people – in an attempt to discover how indigenous people understood and used these species. Much of this information has been erased, or is still viewed through the lens of structural racism. And while the white, male plant hunters who risked their lives to track down new species are celebrated, the indigenous people and enslaved people who carried their bags, cooked their food and shared with them a rich plant knowledge almost always remain nameless: they did not and do not receive the same accolades and financial rewards, and many of them died horrible deaths in pursuit of plants.

I hope that by understanding the histories of houseplants, and recognising the sacrifices of the people who were caught up in the damage done by colonial botany, we can put a different kind of value on the things we grow, and look at plants anew: with respect, deeper insight and an even greater passion. So, put on your 'plant glasses' and let's discover the incredible stories our houseplants have to tell.

A NOTE ABOUT HOUSEPLANTS AS FOOD AND MEDICINE

While many species listed in this book can be classed as toxic, I refer to traditional medicinal and culinary uses for some houseplants. Always seek advice from your doctor before trying any plant as a medicine. And as with any new food, it is advisable to introduce any edible houseplants gradually to your diet. Bear in mind that plants sold as ornamentals may have been treated with leaf shine sprays, insecticides and other chemicals that it would be unwise to ingest, so if you are buying a plant specifically to consume, give it several months to put on new growth before harvesting.

A Note on Houseplant Care

Taking care of houseplants is 90% observation and 10% perspiration. Light should be your prime consideration. Monitor the light in your home: how big are your windows? Which compass point do they face? Does the sun pour in or is it obscured by curtains and blinds inside, or other buildings and trees outside? How does this change with the seasons?

Placing most species on the other side of the room from a small north-facing window is effectively sentencing them to a slow death. Light levels are dramatically lower inside our homes than they are outside, so most plants will need to be close to a window when grown indoors, or have light supplemented by growlights. Examine your plants' leaves to assess how much light they need. Light-reflecting grey hairs, succulence or translucent leaf 'windows' are clues that your plant lives in a place of strong sunlight, so will need your largest, sunniest south-facing window. Leaves that are tissue-thin or have red undersides usually come from places with high humidity and lower light, so will need protection from the midday sun. Any changes to light levels should be made gradually, though: like humans, plants build up their natural protection from the sun's radiation over time, and even a cactus will get sunburned if moved straight to your sunniest windowsill.

Scientific names of plants will offer an indication of their origins, and help you distinguish one species from another. They are split into two parts: the first part (the *Monstera* in *Monstera deliciosa*) is known as the genus – the plant equivalent of a human surname. The second part is the specific epithet – the given name, if you will. Each genus is part of a family, made up of plants with similar characteristics: for instance, the genus Monstera is part of the Araceae, or aroid family. You will find the family of each plant profiled in this book below its common name on the opening page of the chapter. Knowing a plant's genus, species and family will help discover where and how your plant grows in the wild. Is it a denizen of the understorey of tropical rainforests that lives on trees, used to growing in small pockets of fast-draining, humus-rich soil, dappled light and high humidity? Or a desert dweller that grows in cracks in rocky ground and faces daytime temperatures of 40°C (104°F)? This information can guide you on caring for your plant at home. And if you get light levels and potting mix right, watering then becomes a much less tricky proposition. After an initial honeymoon period – which could be as short as a day or as long as a couple of months – new plants start adapting to the conditions in your home, which may be very different from the computer-controlled nirvana of the nursery. Leaves may drop, because the plant is adjusting to lower light conditions. Some plants sent via mail order or left to languish in the shop or garden centre may develop marks from sunburn, cold damage or knocks they have experienced before they came into your care. Most will bounce back, but accept that leaf damage is permanent: unlike human skin, leaves cannot repair themselves, but new foliage should soon arrive. Quarantine new plants well away from the rest of your collection for the first six weeks to allow time for any unwanted hitchhikers such as mealybugs to make themselves known.

If your plant did not come from a specialist grower, it is highly likely it is planted into a potting medium that suits the perfect conditions of the nursery, but is less than ideal for home growers. Should you repot these plants as soon as you get them home? It depends: I usually give new plants a chance to settle down for a few weeks before repotting, but if the plant is already rootbound, with a mass of roots circling the pot, or if the substrate is really poor, I do the job straight away.

Observing plants closely often reveals whether your plant is thriving. I start a plant inspection with the roots. Take the plant out of its container: are the roots firm and sweet smelling? Is the substrate dry as a bone and compacted, or moist and airy? Has the rootball run out of space in the pot? Next, look at the leaves, especially the undersides where pests tend to proliferate. A botanist's hand lens or magnifying glass will help you spot red spider mites, which are too small to see with the naked eye, and will also reveal the juvenile forms of other pests such as scale, mealybugs and thrips. New growth is often a home for pests such as aphids, which love these sap-filled tender parts.

When it comes to pest control, people often look for a 'silver bullet' solution they can apply once and so obliterate the problem. This is very rarely achievable. The simplest and best form of pest control is watching your plants like a hawk and wiping off any pests as soon as they appear with a damp cloth, or washing plants down outside or in the bath or shower. Biological controls – using predatory creatures such as mites or nematodes – can be very effective, and are safe for people and pets, but they do not usually work on every pest. The more you learn about the habits and lifecycles of pests, the better you can address how to control them.

Indoor gardens are never going to be free of living creatures, and not all so-called 'creepy crawlies' are out to do your plants harm. Springtails often live in houseplant soil and get mistaken for some heinous pest. Wrong – they don't harm plants but do eat fungi and decaying vegetation in the soil. Probably the most notorious pest of all is the one that does the least damage to plants: fungus gnats. These tiny floaty black flies lay their eggs in houseplant soil and their larvae feed on fungi and decaying matter. The adults love to waft around your head, lured by the carbon dioxide in human breath. The internet is awash with 'home remedies', but the best treatment is a twice-yearly application of a biological control: either the nematode *Steinernema feltiae,* or hypoaspis mites, *Stratiolaelaps scimitus.*

When it comes to watering, observation is key (you may be beginning to notice a pattern here). Watering to a schedule – 'it's Sunday, so I'll water' – can result in plants getting too much or too little to drink. Instead, find out what is going on at root level. The surface may be dry, but a few centimetres down, the substrate can still be distinctly damp. You can test this in various ways: with your finger, a wooden kebab stick or a moisture meter. When you water, make sure that all the substrate is wetted evenly, and do not let water sit in the base of the pot for more than a few minutes – unless it's one of the few plants that like to sit in water, such as Venus flytraps. I have the luxury of using rainwater from my garden on my plants – this avoids issues with the buildup of mineral salts found in the 'hard' water where I live. Some tap water also contains chlorine and fluoride, which can be problematic for some plants. Most houseplants will be fine with tap water (except for carnivorous plants, which really do need rainwater or distilled water), but if you can collect some rainwater, do use it on your plants.

When it comes to choosing a substrate, look for a peat-free mix. Peat bogs are vital carbon sinks and biodiversity hotspots, so there is no justification for tearing peat from the ground to use in horticulture, and there are many viable alternatives. Every grower has their own unique recipe for their plants' substrates, adding a range of materials such as perlite, vermiculite, pumice, sand, grit and pine bark to improve drainage, aid water retention or add more air pockets to the soil. Some prefer to grow in an inorganic mix such as pon (a soil-free, mineral-based substrate) or leca (expanded clay pebbles), employing hydroponic or semi-hydroponic techniques. Whatever approach you choose, remember that roots crave the same things: air, water and nutrients. The many houseplant species that grow as epiphytes, including Hoyas, bromeliads and Peperomias, have small rootballs and prefer airy soil. Cacti and succulents often dissolve into mush in the hands of a beginner grower because they have been planted into a potting mix that holds onto too much moisture, so use a specialist cactus and succulent substrate that includes plenty of drainage material.

And do not forget to feed your plants. My general rule is to feed a dilute solution of houseplant fertiliser regularly in the water whenever I can see they are actively growing. Plants newly repotted in a soil-based mix (as opposed to pure pon, pumice or leca) will be supplied with enough nutrients to last several weeks, so you can hold off feeding until that time has elapsed.

If your head is now spinning, remember: experimentation is how you learn about what plants thrive best and where in the unique conditions of your home. You may kill some plants along the way, and that is fine. After many years of growing, I still kill plants too. Arm yourself with as much knowledge as you can, grow things that make your heart skip a beat, and the rest will follow.

Aegagropila linnaei

MARIMO MOSS BALL

Pithophoraceae

L et me start with the bad news: marimo moss balls are neither moss nor plants. 'Filamentous freshwater green macroalgae', while more accurate, does not have the same ring to it.

So, what are algae doing in a houseplant book? Marimo balls have been popular in the aquarium trade for many years, where they were sold as curious greenery for freshwater tanks, but more recently they have been marketed anew as the ultimate low-maintenance houseplant. Marimo may lack the thrills and spills of the vascular plants with actual leaves and flowers we grow indoors – but you will have a tough time killing one.

Perhaps I am being unfair. Marimo balls have tapped into a desire for an undemanding, furry 'pet' that won't mind if you forget to care for it for weeks while offering a low-tech alternative to the flurry of mindfulness apps. Drop a marimo or two into a clear glass jar of water and this emerald green, velvety ball will slowly rise and fall over the course of a few days, rather like a living lava lamp. The specimens you're most likely to find for sale will nestle in the palm of your hand, but they can reach 30–40cm across, given time.

With an annual growth rate of about 5mm, larger balls some-times found in the wild will be decades or even centuries old. You're unlikely to spot one of these ancient marimo balls, though, because they dwell at the bottom of lakes. They are made of a species of algae called *Aegagropila linnaei* found in freshwater and brackish lakes and some rivers in several countries in the northern hemisphere, including Scotland, Ireland, Ukraine, Estonia and Iceland. This species doesn't always grow in balls; indeed, the spherical form is a rarity. It can more commonly be found growing as a mat made up of numerous filaments, either floating free or attached to rocks or the floor of the lake, or as a mass of unattached filaments free-ranging through the water. Even then, it exists in fewer locations across the northern hemisphere than other algae species. One of the reasons why *A. linnaei* is not more widespread relates to its lack of akinetes – thick-walled, dormant cells that are resistant to drying out. When other species of algae are picked up on the beaks of water birds and deposited later in another body of water, these akinetes spring to life to create new algae.

So, why is this particular species sometimes found as a ball? Some have theorised that this shape made it harder for fish to eat them – although anyone who has seen a marimo torn apart by a goldfish may dispute this. Scientists' most convincing theory is that fragments of the algae in mat form break loose and get rolled around by the waves and currents in shallow water, slowly aggregating into a spherical shape, allowing light to enter from all angles and for sediment to be displaced as they are rocked around by the movement of the water on the soft, sandy lake bottom. These can accumulate into natural 'ball pits' up to half a metre deep when conditions are right. Scientists have discovered that these layers of balls can achieve greater biomass – and most likely productivity – than a simple mat.

Smaller balls are solid all the way through, although once they exceed 10cm across, they begin to develop a hollow inside, as light cannot reach deeper than 5cm into the centre to allow photosynthesis. Why do marimo balls float up and down? This isn't fully understood, although the basics are clear. By rising from the lake bottom to the surface, marimo can access more light for photosynthesis. Bubbles of oxygen created by photosynthesis coalesce on the upper surface and add buoyancy, allowing it to float to the surface. This seems to follow circadian rhythms, and may have applications for humans too. As fanciful as it may sound, scientists have even begun to create prototypes of 'biomimetic devices' which emulate or draw inspiration from nature, including motors powered by the movement of marimo balls.

Despite the presence of marimo in several countries – the vast majority of plants in the nursery trade originate in Ukraine – they are most famous for growing in Lake Akan in Hokkaido, Japan. The name marimo was devised by a Japanese botantist called Tatsuhiko Kawakami – a portmanteau word joining *mari*, meaning bouncy ball, to *mo*, meaning aquatic plant (yes, language fans, 'marimo ball' is a tautology). Japan designated marimo as a national treasure a century ago, and in 1952 they were further promoted to the status of special natural monument.

Yet the existence of marimo is still threatened by human development, even as they have become a popular product in the aquatic and houseplant industries. *Aegagropila linnaei* is now extinct in some places, including most sites where it had been found in the Netherlands. Marimo thrive in lakes with low nutrient levels, but populations have been damaged by pollution, souvenir hunting, removal for the plant trade and runoff of sediment, and nitrogen and phosphorus caused by deforestation and development.

The indigenous Ainu people who live around Lake Akan call marimo *torasample* (lake goblin) and *tokarip* (lake roller). In 1950, realising that this national treasure was rapidly disappearing from the lake, they came up with a way of trying to preserve marimo and encourage those who had taken balls as souvenirs to return them by honouring the algae with a festival held every October. From torchlit processions to dance demonstrations, the Marimo Matsuri culminates in the return of symbolic marimo to the lake. In the last few decades, marimo has spawned a whole range of cultural products, including several mascots. The most famous is the Hokkaido mascot Marimokkori, another portmanteau word where *mokkori* is variously translated as erection or bulge. This curious character has a marimo for a head, and another as a crotch or navel, depending on your mindset. There's even a pullstring toy Marimokkori, where the marimo bulge can be pulled to make the character speak.

Marimo may be the world's most charismatic algae, but its rise to fame in the houseplant world may be cut short. Their sale has been outlawed in some parts of the world, including Tasmania and New Zealand, because they pose a risk as an invasive species. In 2021, zebra mussels (*Dreissena polymorpha*) were found on a marimo in a Seattle pet store, and subsequently found in at least 20 other US states. This was disastrous news, as these molluscs are a highly invasive species that threatens waterways around the world if allowed to spread. The US Fish and Wildlife Service urged Americans to safely destroy marimo bought after 1 February 2021.

CARE GUIDE

Light When you first add marimo to water, they often float because they are holding onto lots of air bubbles. They usually sink after a few hours or days. Depending on light conditions, they may rise and fall, or stay on the bottom. If your balls aren't moving, try increasing their light exposure, especially in the morning. Keep the container away from direct sun.

Temperature Marimo will thrive in room temperature water, and in freshwater fish tanks either unheated or heated for tropical fish, but will struggle above 26°C (79°F) so aren't suitable for tanks heated to above this temperature. Some people temporarily remove marimo to a cooler spot or even to the fridge during hot weather.

Water Marimo must be immersed in water to survive, although they can cope with a few hours or days out of water if kept damp. Use tap water or reverse osmosis water to fill up your container: if your tap water is heavily treated, it's worth treating it with dechlorinator used to prep water for an aquarium. Do not use rainwater, as this can introduce pathogens and unwanted creatures such as mosquito larvae to your container. Sometimes adding carbonated water is recommended as a way of perking up poorly marimo, but this is unnecessary (and wasteful). Changing the water more regularly is far more effective. Marimo can grow in brackish water, so adding a pinch of sea salt or aquarium salt to the water is recommended by some, but it is not essential.

Humidity Given marimo balls grow underwater, this is not a concern.

Pests and diseases Fish tank marimo may become home to snails, but these won't do any harm. See the note about zebra mussels above.

Substrate Marimo balls can just sit in water, but include a bed of gravel or decorative stones if you wish.

Propagation You can split a larger ball by squeezing it to ring out excess water, then tearing or cutting it into smaller pieces. Roll these between your palms to create smaller balls, but bear in mind their growth rate will remain a leisurely 5mm a year. Some people tie fishing wire or cotton around newly formed balls to help them adhere.

Feeding Unnecessary.

Other maintenance tasks
When you change the water, gently squeeze marimo while rinsing in water to remove any debris that has collected on the surface. Rotate balls regularly so that all sides can receive light.

Danger signs Marimo should be a velvety dark green, and shouldn't smell bad. Balls turning white may be receiving too much sun, while patches of grey may indicate the marimo lacks light. Balls turning black are dying. Marimo growing unevenly should be rotated more frequently.

Toxicity No known issues. *Aegagropila linnaei* has been found to be a component of the food product *kai*, collected in northern Thailand, so seems to be edible, but I would not recommend munching on your marimo.

Display Any clear glass container is fine as long as the marimo are completely covered. Add a lid if you want to keep out children or pets, but be especially sure to change the water every few weeks and take the lid off and give the water a swirl occasionally, especially if the container is deep, to keep the water oxygenated. Marimo do well in tanks, either on their own or as part of a planted display with or without fish. Avoid cohabiting them with fish or other creatures that will destroy them: crayfish and goldfish, for example, often tear marimo apart.

Cultivars None.

Also try ... There are many aquatic or semi-aquatic plants that make small but interesting subjects for growing in water or moist terrariums, such as Cryptocoryne and Anubias from the aroid family.

Aloe vera

TRUE ALOE

Asphodelaceae

*A**loe vera* is the best known of more than 500 species in the genus, but physically at least, it is far from the most interesting: at 3m or more tall, the bitter aloe, *A. ferox*, is far more imposing; *A. variegata*, the tiger aloe, has prettier leaves; and the flowers of the prickly aloe, *A. aculeata*, are more dramatic. But of all the species profiled in this book, *A. vera* is the only plant that does double duty as a multi-billion-pound commercial crop and a ubiquitous, much-loved houseplant. And throughout thousands of years of cultivation, *A. vera* has gathered a huge freight of folklore and claims about its powers – medicinal, nutritional and even spiritual. Yet much about it remains a mystery.

Aloe vera is one of the stemless, rosette-forming Aloe species, and can grow waist-high in the right soil and climate, although homegrown specimens are more likely to be 30–50cm tall. These are generous plants, throwing out lots of suckers that produce young plants or 'pups' around the base of the rosette, which can be pulled away to make new plants. If you don't own one, ask around and you should be able to find friends or family who will gift you a pup.

When young, the toothed leaves are stippled with silvery marks, which disappear as the plant matures and the leaves grow bigger and more leathery. Botanists haven't yet put their finger on the exact reason for this variegation, which is common to many Aloe species, but Colin Walker, co-author of *Aloes: The Definitive Guide*, speculates it may be a strategy to deter herbivores from munching on the young leaves. Grown outdoors, *A. vera* produces tubular, acid-yellow flowers clustered together on stems that hold them aloft from the leaves, but it rarely blooms when grown indoors in a pot. Unlike some other rosette-forming succulents such as

Sempervivums and most of the Agaves, the plants do not die after flowering.

One place where you will see *A. vera* in flower is on farms in tropical regions around the world, where it is grown by the million to feed demand as an ingredient in cosmetics, toiletries and food-stuffs. It features in a mind-boggling array of products, from soaps and moisturisers to drinks and yoghurts, and it's also sold dried, powdered and as fresh leaves for home processing.

This succulent has been cultivated for so many centuries and become naturalised in so many subtropical and tropical countries, from India to South America, that many nations have claimed *A. vera* as their own. But its precise ancestral origins have, until very recently at least, remained unclear. Aloe species are found in a range of locations across sub-Saharan Africa, Madagascar and the Arabian peninsula. In 2015, a team of scientists led by Olwen Grace of the Royal Botanic Gardens at Kew and Nina Rønsted of the University of Copenhagen finally provided an answer. DNA analysis of *A. vera* and other Aloe species allowed Grace and colleagues to narrow down its homelands to the Al-Hajar mountains of north-eastern Oman, in the Arabian peninsula.

Starting to solve the genetic puzzle of *A. vera*'s ancestry helps to explain its physical characteristics: it evolved to survive the dry climate and rocky, nutrient-poor soils of an arid, mountainous land-scape by combining a leathery skin – known as a waxy cuticle – on the outside of the leaves with a fleshy interior. The carbohydrate-loaded tissue that makes up the fleshy part stores water and nutrients the plant can draw on during droughts. This translucent, gel-like sub-stance is cut away from the outer waxy leaf cuticle when the leaves are processed for the commercial market, in a movement similar to filleting a fish. It's this mucilaginous substance that has earned this species the common name *erva-babosa* in Brazilian Portuguese, which means 'slobber herb'.

That is not *A. vera*'s only product: a yellow liquid, known as exudate, can be extracted from just below the surface of the leaf. I will not attempt to evaluate the vast and sometimes contra-dictory scientific literature on the effectiveness of *A. vera* gel and exudate for a bewilderingly huge array of conditions suffered by both humans and livestock, including baldness, headaches, burns, dysentery, stomach ulcers, digestive problems, parasite infestations, fever convulsions, radiation burns, genital sores, skin inflammation and cancer. My own use of *A. vera* is limited to grabbing a leaf and applying the gel to the skin to treat minor burns.

I am even less qualified to validate *A. vera*'s other application, as a talisman reputed to bring good luck and fend off evil. I came across numerous examples of this practice across various cultures and time periods, with a concentration in South America and Egypt. Egyptologist Edward William Lane reported in 1825 the 'very common custom' of hanging an Aloe over the doorways of homes 'as a charm to ensure long and flourishing lives to the inmates, and long continuance to the house itself'. In 2015, anthropologist Nicholas C. Kawa saw an *A. vera* hanging in a shop in Iquitos, Peru. When asked, the bodega owner told him, 'It's to keep the bad vibes (*malas vibras*) away.' A few years earlier, ethnobotanists reported in the *Journal of Ethnopharmacology* that Spanish-speaking Latino immigrants in London grew *A. vera* in pots by their doors for the same reason. It has also been used as a boundary marker and a grave plant in the Middle East: Gilbert Reynolds' 1966 book *The Aloes of Tropical Africa and Madagascar* records that Aloe is grown in Egypt 'especially as a cemetery plant, and sometimes as boundary marks demarcating fields'.

To delve into the first documented references to this plant, we need to go back to the first century CE, when Greek physician and botanist Pedanius Dioscorides wrote *De Materia Medica*, his five-volume work on medicinal plants. The original document has been lost to history, but a parchment dating to 512 CE known as the *Codex Aniciae julianae* contains the first known image of *A. vera*. It looks remarkably like the one I have in my sunroom, barring the fat flower spike. There are many claims for *A. vera*'s fame even before the sixth century, including impossible-to-verify stories that Cleopatra used the plant in her beauty regimen, and the ancient Egyptians used it as part of the process of embalming bodies. Was this *A. vera*, or some other member of the Aloes? We may never know, but the cumulative historic power of the *A. vera* story has cemented this plant as deserving of the title of true aloe.

Does *A. vera* possess any qualities or ingredients that render it more powerful than any other Aloe? Olwen Grace and the other scientists who located its origins believe its preeminence is more a case of geography: true aloe originated close to important historic trade routes, and came into cultivation earlier than other species with similar qualities. Some Aloe experts even postulate that *A. vera*'s reluctance to flower and eagerness to send out suckers to create new plants may indicate that this species is in fact a hybrid: a cross between two Aloe species dating so far back that its parentage has been lost.

CARE GUIDE

Light Your sunniest windowsill is ideal for this plant, although it sustains itself in a gloomy corner without keeling over provided you get watering and substrate right (see below). The grey-green leaves sometimes take on a reddish tinge if underwatered or suddenly exposed to more light than they're used to. If your plant gets leggy, it's probably not getting enough light.

Temperature It will take as much heat as you can throw at it in summer. But *A. vera* needs a dormant period in winter, brought on by lowering temperatures to around 10–15°C (50–59°F) and allowing the potting mix to dry out almost completely.

Water During the growing season, water generously. From November to March, leave the substrate just shy of dry, particularly if the plant is being kept cool.

Humidity Perfectly adapted to cope with dry air.

Pests and diseases Mealybugs are probably its number one enemy.

Substrate *Aloe vera* roots are adapted to live in rocky, free-draining ground. As houseplants, they benefit from a potting mix containing half drainage material (perlite, grit) and half a regular houseplant mix such as John Innes No. 2.

Propagation It's not possible to propagate *A. vera* from an individual leaf. Nevertheless, propagation is easy: plants produce numerous pups or young plants from around the base of the rosette. Carefully tease these away and repot separately in gritty potting mix.

Feeding Use a half strength houseplant feed or specialist cactus and succulent feed occasionally when the plant is in active growth.

Other maintenance tasks Pull away dead leaves from the base only when they are completely desiccated.

Danger signs Leaves may become soft and puckered when the plant is left without water for long stretches, although similar symptoms can show signs of root rot caused by overwatering, so check the rootball before watering. Weak, spindly growth often occurs when plants do not receive enough sunlight. Plants may turn pale if exposed to low temperatures, or if suddenly moved from shade to bright sunlight.

Toxicity Toxic to pets; while *A. vera* extracts are present in many dietary supplements and other products, research is ongoing as to their possible health benefits and risks to humans. Most experts agree that using *A. vera* on the skin is safe, but seek medical advice before taking it internally.

Display Planting in a terracotta pot will allow more air to access the roots; if you use plastic or glazed china, it's all the more important to ensure your potting mix is free draining. Leaves that rest on the edges of a terracotta pot tend to rot. It's an aesthetic decision whether you prefer your *A. vera* to stand alone as an architectural feature, or to leave pups to proliferate and create a clump.

Cultivars None.

Also try . . . There are many Aloe species and hybrids that make wonderful houseplants: the many-leaved aloe, *A. polyphylla*, is worth growing for its fractal-shaped rosettes, while the tiger tooth aloe, *A. juvenna*, is compact enough for the narrowest windowsill.

Aspidistra elatior

CAST IRON PLANT

Asparagaceae

Readers perusing the small ads in *The Times c.*1932 in search of a holiday getaway may well have come across a small advertisement for Tower House. This hotel boasted of being 'the warmest house in Bournemouth' (a resort town on England's south coast) and mentioned its spacious lounges and bedrooms with hot and cold water, concluding with the 1930s equivalent of a mic drop – '(NO ASPIDISTRAS)'.

How could a plant be so reviled that an English guesthouse felt obliged to assure guests of an Aspidistra-free environment? The fortunes of the cast iron plant have waxed and waned dramatically since the Japanese species was first described to science in 1834 by the German–Dutch botanist and plant collector Carl Ludwig Blume. The Dutch had exclusive trading rights with Japan, hence it was they who first introduced this plant into Europe. In 1861 the *Gardeners' Chronicle* reported that Belgian horticulturist Louis Van Houtte recommended the 'little known plant' for growing in sitting rooms due to its ability to survive shade and fluctuations in temperature. Thirty years later, the Aspidistra topped a list of the twenty best plants for living rooms in the same publication. It spread across the Atlantic just as speedily, becoming a fixture of American parlours, and being adopted as an outdoor bedding plant in more southerly states where it could survive the winter. Some of its array of other common names – bar-room plant and spittoon plant in the UK and the US, and the butcher's plant and cobbler's palm in the Netherlands and Germany – further reflect the Aspidistra's ability to survive in far from ideal conditions.

By 1900, no parlour in a middle-class home was complete without a cast iron plant, a living symbol of Victorian respectability.

There is a story that every Aspidistra leaf symbolised a hundred pounds of a family's annual income. True or not, this was a plant that people loved to put on show, and many photographs still exist of Victorians posing alongside their Aspidistras. Plants were usually nestled in a jardiniere and framed by curtains that allowed passers-by ample views of the Aspidistra, but kept the rest of the room obscured from view.

To what can we ascribe the Aspidistra's rise to Victorian house-plant ascendancy? This species has a hidden superpower – its ability to withstand the ethylene-laced air that occupied Victorian homes lit by gas and heated by coal fires. You will have harnessed the effects of ethylene if you've ever put a banana in a bag to hasten the ripening of green tomatoes; it's also the plant hormone used to initiate flowering in bromeliads, including the pineapple (*Ananas comosus*). But when plants were exposed to the levels of ethylene given off by Victorian-era gas lights, most would struggle, become leggy, lose leaves and flowers and eventually die. *Aspidistra elatior* is able to cope with far higher ethylene levels than other house-plants (aside from fellow Victorian favourite *Howea forsteriana*, see page 99); it may have become such a staple because it simply survived where other plants did not.

Even once gas lighting began to be replaced by electric lights in the early twentieth century, the reign of the Aspidistra was not over. Things started to look shaky in the 1920s, when fussy, overpacked Victorian-era interiors began to be emptied and replaced with the clean lines of Art Deco. Aspidistras became a symbol of everything people wanted to forget about the previous century. In 1926, a play called *Aspidistras* by Joan Temple received a lukewarm reception in *The Times*, with the reviewer noting: 'The aspidistral background, decorated lavishly with the pale paper flowers of suburban humour, has been used so often that a dramatic of some importance is needed to justify the use of it again.'

The Aspidistra was thus well cemented as the butt of jokes by the time George Orwell's novel *Keep the Aspidistra Flying* was published in 1936. In this semi-autobiographical work, he poured all his contempt for a middle-class money-grubbing mentality into the Aspidistra. Orwell's main character, a hard-up advertising copy-writer named Gordon Comstock, finds himself renting a dingy room from a landlady who has thoughtfully provided an Aspidistra in an effort to make him feel at home. Comstock does his best to kill it, failing because, of course, 'the beastly things are practically immortal'. Later in the novel, Orwell writes: 'The Aspidistra, flower of England! It ought to be on our coat of arms instead of the lion and the unicorn. There will be no revolution in England while there are Aspidistras in the windows.'

English music hall star Gracie Fields' song 'The Biggest Aspidistra in the World' came out two years later, and was much more playful than Orwell's polemic. The lyrics tell of an Aspidistra that gets (improbably, admittedly) crossed with an oak tree and ends up so large it has to be watered by the fire brigade. The song was popular enough to be reworked in 1941 to include a line about stringing Hitler from the highest branch. The same year, Prime Minister Winston Churchill gave the name Aspidistra to a powerful radio transmitter that broadcast black propaganda to the German people.

The Aspidistra may have been officially out of fashion, but people continued to place their plants in pride of place for decades to come. My favourite Aspidistra photograph dates from 1940, and shows the aftermath of an air raid on London. Two women stand in front of the brick skeleton of a terraced house, knee deep in masonry, a scene of utter devastation all around. One woman holds an Aspidistra in a jardiniere, while the other grasps a smaller Aspidistra (perhaps a division of the first?) in a terracotta pot with one hand and a mantel clock with the other. At a time when all other points of reference seemed lost, the solid, durable Aspidistra remained: you couldn't even blow it up.

Let's leave behind the Aspidistras of the past to consider how the plant grows in its indigenous home, on the islands of southern Japan. The Aspidistra genus contains around 130 species, many of which have only been found in the last few decades: the discovery of Aspidistra species similar to *A. elatior* in central Vietnam and Taiwan has muddied the taxonomic waters for this species, not helped by a historic rumour – unfounded as far as scientists know – that *A. elatior* originated in China, like many other species in the same genus.

The climate of the southern Japanese islands is humid and subtropical, and Aspidistras grow as an understorey plant, living under the heavy shade of a canopy of broadleaved evergreen trees including the Japanese oak (*Quercus acuta*), itajii chinkapin

15

(*Castanopsis sieboldii*) and the smaller shrub *Ardisia sieboldii*. Temperatures rarely venture below freezing in its homeland, but *A. elatior* has been found to be tolerant to at least −15°C (5°F) and its underground parts will survive at −20°C (−4°F) or even lower. Provided it is grown in shade, it can also cope with baking heat. It grows slowly around tree roots, gradually expanding via an underground rhizome that can store water and nutrients during drought. This rhizome earned the plant the name *chu-ken ch'i*, meaning ground centipede in Chinese medicine, where it is used for a range of conditions including diarrhoea, sore muscles and abdominal cramps.

Rummage around the base of a clump of Aspidistra from March onwards and you may spot its other secret. Aspidistra flowers are described by scientists as cryptic, because their curious flesh-coloured blooms are half-buried in the soil and leaf litter. Botanists theorised that slugs were the pollinators of these ground-level flowers, but Japanese scientists recently discovered that the main pollinator is in fact that sworn enemy of the houseplant grower, the fungus gnat. The Aspidistra's pollination strategy is to mimic a mushroom in both smell and looks. The fungus gnats are drawn in, thinking this would make a great place to lay their eggs, but end up covered in pollen instead, which is useless to them, but highly advantageous to the Aspidistra. The structure of the flower forces the tiny flies to squeeze past its relatively enormous stigma – the female sex organ – to reach the male parts below where they can access the plant's pollen via the anthers. They squeeze back out, and on visiting the next flower, they end up brushing pollen onto its stigma. I hope this knowledge gives you a thrill of excitement if your plant decides to bloom: and who doesn't have a few fungus gnats wafting around ready to serve as pollinators?

Back to the Aspidistra's more obvious assets: the leathery leaves. These have become a fixture of flower arrangers around the world, prized for their durability and flexibility, the perfect foil for floral displays. The Japanese floral art form ikebana often incorporates Aspidistra leaves, which are said to symbolise improving fortunes. One of the first skills ikebana trainees learn is how to split, curl, twist and otherwise manipulate Aspidistra foliage. However, the leaves are not symmetrical: one side is always wider than the other, usually the right, so finding 'left-handed' leaves is a challenge to ikebana practitioners.

Japanese chefs have also made good use of Aspidistra foliage. The bright green tufts of plastic found adorning bento boxes or sushi today are an echo of a much older tradition that faded from use in the 1960s: using plant leaves to keep food flavours distinct by separating rice from fish. Aspidistra was usually the leaf of choice in Kyoto and Osaka, whereas chefs in Tokyo tended to use Japanese bamboo leaves (*Sasa veitchii*). Some chefs are returning to the use of real Aspidistra, known in Japanese as *baran* or *haran*, as a way of cutting down on plastic, but there is another benefit too. Aspidistra contains high levels of phytoncides, antimicrobial chemical compounds that inhibit growth of bacteria and fungi, thus helping to keep the food fresher for longer.

A combination of *A. elatior*'s imperishable nature and its ability to pump out phytoncides to reduce airborne moulds and bacteria as well as remove polluting gases such as carbon dioxide and benzene has resulted in a new role for the species, as an air purifying plant for homes and offices. While these claims may be overblown – you'd need to pack your rooms with plants for any significant effect and you're far better off just opening a window – it's fascinating to watch the wheel of fortune spin once again for this most indomitable of houseplants.

CARE GUIDE

Light Even an Aspidistra won't do well in the darkest recesses of a room with a small north-facing window. But it is still one of the most shade-tolerant houseplants there is. The leaves will quickly burn in direct sunlight, but it can tolerate a bright spot.

Temperature *Aspidistra elatior* is supremely tolerant of temperature variations, down to −15°C (5°F) and even up to 50°C (122°F) or more (if shaded).

Water In winter, allow the substrate to dry out so you can feel dryness at root level before watering. The rest of the year, water when the surface is dry. Aspidistra roots will rot if left sitting in water, so take care to drain well after watering.

Humidity Aspidistras cope perfectly well with the low humidity of many modern homes.

Pests and diseases Scale and spider mite are the main enemies of indoor Aspidistras, but they can succumb to thrips and mealybugs. Slugs and snails may hitchhike in on plants left outside for the summer, so check underneath pots before bringing them inside. Another issue with plants that spend time outside may be vine weevil, which eat away irregular notches on the leaf edges.

Substrate Many Aspidistras will go years without being repotted, and may sulk after transfer; around every three years should be acceptable for most plants. They are tolerant of a wide range of potting mixes, but John Innes No. 2 or 3, peat-free houseplant substrate, or any regular foliage houseplant substrate will work.

Propagation Cut the rhizome into pieces, making sure each section includes a growth point, and pot up separately.

Feeding Aspidistras have low nutrient needs, so feed with a dilute fertiliser for foliage houseplants once a month when the plant is in active growth.

Other maintenance tasks Cleaning leaves is always mentioned as the main maintenance task, as there isn't much else to caring for this plant.

Danger signs Pale stippled leaf undersides usually indicate mites or thrips. Variegated cultivars sometimes turn plain green when grown indoors, a sign that they are not receiving enough light: a brighter spot indoors or a spell outside usually helps produce new leaves with patterning.

Toxicity No known issues.

Display To get the Victorian parlour look, display your Aspidistra in an antique spittoon or jardiniere: pillar-shaped plant stands are perfect for large specimens. Aspidistras also look good in a simple terracotta pot.

Cultivars There are several variegated cultivars of *A. elatior* including 'Okame' with cream-striped leaves, spotty 'Hoshi-Zora' and the striking 'Asahi', whose leaf tips are daubed with cream.

Also try . . . There are many other delights to enjoy within the genus. Many have elaborately spotted, speckled or striped leaves; my favourite, *A. zongbayi* 'Uan Fat Lady', which is indigenous to China, has all three.

Begonia maculata

POLKA DOT BEGONIA

Begoniaceae

Social media filters have got a lot to answer for, but one of their most heinous crimes is misrepresenting the colour of *Begonia maculata* leaves. I exaggerate. But scroll Instagram and you can be forgiven for thinking that the angel-wing foliage of this Begonia is one shade of green away from black. This illusion creates a vivid contrast to the plant's silvery splotches, but it is not real: the green of *B. maculata* leaves is the olive drab of an army surplus parka, not the deep green of an avocado skin.

The levels of adoration received by *B. maculata* – and the cultivar 'Wightii' in particular – in the last few years are astonishing: #begoniamaculata is more popular than any other species-specific Begonia hashtag on Instagram, with #begoniarex in second place, garnering just over half the number of mentions. And yet this was not always so: scanning through houseplant books and articles for mentions of the polka dot begonia before this boom, it does not stand out as the most iconic species. The *California Horticulturist* in 1871 listed *B. maculata* among a dozen or so of the best species to be obtained at the time, noting its leaves as being 'somewhat peculiar in shape' though 'elegantly marked with white spots'. Twenty years later, the *Gardening World Illustrated* damned *B. maculata* with faint praise, recommending it for growing in windows in Manchester in northern England because it 'stands washing well when dusty' – admittedly a quality much needed in a plant grown amid a period of heavy pollution.

Begonia maculata is native to south-eastern Brazil; it grows as an understorey plant in forests, and has become naturalised in other Central and South American countries including Cuba and Argentina. Its type locality – the place where the first specimens described by botanists were found – was a mountain called Corcovado in central Rio de Janeiro. Corcovado is most famous for the 30m tall statue at its peak, *Christ the Redeemer*. Italian botanist Giuseppe Raddi named and described the species for western science in 1820, and it is presumed this is one of the 4,000 specimens he brought back from a seven-month trip to Brazil in 1817.

What is it about this plant that has elevated it to the status of a houseplant icon now? The angel-wing leaves fill the square of an Instagram photo so beautifully, even when not darkened with filters. The silver markings against an olive background, set off by flashes of the cherry red undersides of the leaves, have proved a potent image. As the magazine *Real Simple* put it in 2021, this plant 'looked too cute to be real'. I would, however, urge you to treat with extreme scepticism claims parroted by plant blogs that the leaves' red undersides inspired luxury shoe designer Christian Louboutin to create his signature red sole. I found absolutely no evidence for this, and Louboutin has given many interviews in which he offers another explanation for their origin: he saw his assistant painting her nails with bright red polish. Nevertheless, an association with high end fashion has not done the plant's cachet any harm.

Strangely, the term 'polka dot' was coined decades after Raddi first saw *B. maculata* on a Brazilian mountainside. A craze for an energetic dance called the polka swept the world in the mid-1800s, and somehow this became associated with spotty fabrics. I could not find records of *B. maculata* being called the polka dot plant before the twenty-first century. The common names of trout begonia, fish begonia (presumably because the silvery spots resemble scales) and less commonly clown begonia seem to have dominated as monikers for *B. maculata* before the twenty-first century. There were references to 'polka dot begonia' in some American newspapers in the 1970s, but, with no reference to a scientific name, they may have been referring to one of the many other dotty-leafed Begonia species. The species that has historically been named for its polka dots was another houseplant, *Hypoestes phyllostachya*, with its pink-splashed leaves.

What purpose do spots serve this Begonia? There are a couple of scientific trains of thought and, as is often the case, more than one theory may be true. It is clear how the spots are formed, at least: the areas of the leaves with silver spots contain air spaces between the epidermis, or the upper leaf surface, and the chlorenchyma, the cells that contain chlorophyll and perform photosynthesis. These empty spaces bounce light back out of the plant, giving them a white or silvery look. This kind of variegation is known as blister or air space variegation, and is also found on other houseplant species, including *Hoya carnosa* (see page 107). Blister variegation does not seem to cost the plant in terms of photosynthetic efficiency: chlorophyll levels are unaffected, unlike other forms of variegation where chlorophyll is either reduced or completely absent in areas of the leaf, showing up as paler green, creamy or white patterns.

One theory is that the spots mimic damage from pests such as leaf miners, whose larvae eat away tissue from within, creating silvery tunnels. Researchers have also found that butterflies are put off laying eggs on leaf surfaces that look as if they contain patches of eggs. Could the markings on *B. maculata* serve such a purpose? Possibly. There are also suggestions that these intercellular spaces that create the silvery spots could help to bounce light around inside the leaf and reduce damage from the variations in the intensity of the sun that affect plants living under dappled canopies. The red leaf undersides sported by *B. maculata* are also a fairly common phenomenon of Begonias and other tropical plants that grow in the understorey; but the reason for them remains unclear. Scientists thought the red anthocyanin pigments help to bounce light back up into the cells that perform photosynthesis, to maximise efficiency, but the most recent research suggests that these anthocyanins act as a sunscreen, protecting plants from bright spots of light that penetrate the canopy above.

Begonias are a large genus of around 2,000 species, spread across subtropical and tropical regions, including Central and South America, Africa and South-east Asia. Horticulturists have categorised the species grown in cultivation into eight groups according to the way they grow and are grown: *B. maculata* is classed as a cane begonia. It is a variable species, sometimes known as a 'species complex': some leaves display a strong pattern of spots, while others are plain green. There are several cultivars, the most popular being 'Wightii' (also sometimes called 'Wrightii'), which has pronounced silvery dots that tend not to extend to the edges of the leaf, and vary in size from a couple of millimetres wide to the diameter of a pea. The leaf itself is narrower than in the species, giving the foliage a svelte look. Begonia expert Claude Barrère notes that 'Wightii' is now considered to be a variety, cultivar or even a hybrid that dates back to the early twentieth century. Who or what was Wight or Wright? Despite my best efforts at Begonia sleuthing, the answer appears to be lost.

Rarely are *B. maculata*'s flowers pictured on social media: mainly, I suspect, because many people photograph their newly bought, immature plant, which promptly loses all its leaves and never gets to flower. And yet the polka dot begonia's pink (or in the case of 'Wightii', pure white) flowers, produced from early summer to autumn, are plentiful and attractive. Like other Begonia species, they have separate male and female forms, produced on the same plant: if you look carefully, the male flowers have petals of unequal size, and tend to be less numerous than the female flowers, whose petals are equal. Begonias are self-fertile, meaning pollen can be passed from the male to the female flower on the same plant to produce seeds. *Begonia maculata* has also played parent to many Begonia hybrids.

Several Begonia species have many applications in traditional medicine and as an ingredient in the places where they grow wild, in particular *B. gracilis* of northern Mexico, whose sour stems are rich in oxalic acid and are chewed while the roots are used as a purgative. But I found no indication of *B. maculata* having any other recorded purpose than as an ornamental, so you may enjoy your plant without fear that you are missing out on anything. I do hope, though, we can find more adventurous ways to display this beautiful plant, not least using the foliage in floristry. In 1897, the publication *American Florist* suggested a wreath using *B. maculata* and the brittle maidenhair fern, *Adiantum tenerum* 'Farleyense'. Perhaps a brave reader can pick up this gauntlet and have a go at creating a twenty-first century Begonia wreath.

CARE GUIDE

Light As gorgeous as it may look on your mantelpiece far away from a window, this and other cane begonias will be happy receiving some direct sun, particularly morning sun which is not so intense. Mine does well in my north-facing glass-roofed room, or try an east-facing windowsill.

Temperature It can cope with a minimum of 13°C (55°F) up to 35°C (95°F), but keep it out of cold draughts. When light levels are lower in winter, it will benefit from being kept in cooler temperatures.

Water Watering the 'right' amount is the great challenge that has been the downfall of many a new Begonia grower. The key is choosing the correct substrate, along with a pot with plenty of drainage holes: once that has been put in place, watering Begonias generously helps to keep plants from becoming crispy without leading to root rot. Begonias can be sensitive to fluoride and chlorine in tap water, so if your water contains these, use rainwater, or water from a reverse osmosis system. Once plants are mature, they are much less likely to be troubled by a slightly patchy watering regime; plug plants,

freshly rooted cuttings and young, newly purchased plants are most likely to suffer a sudden collapse.

Humidity This is not a Begonia species that must be kept in the moist fug of a terrarium to survive: it will do well in 50% humidity.

Pests and diseases Begonias are prone to powdery mildew, which can be a difficult disease to treat: good air circulation and overall plant health will help.

Substrate *Begonia maculata*'s root system is fibrous and fairly shallow, so the substrate needs to be airy. Their worst enemy is a substrate that holds too much water and causes root rot, which results in rapid leaf drop. Every Begonia grower has their own unique recipe, but mixing a decent amount of drainage material such as perlite or leca – at least a third – into a regular houseplant potting mix should work well for this species.

Propagation *Begonia maculata* is relatively easy to propagate from stem cuttings: place in a glass of water, a pot of Begonia substrate or a mix of sphagnum moss and perlite, sealing both pot and cutting in a clear plastic bag.

Feeding Feed with foliage houseplant fertiliser when the plant is in active growth.

Other maintenance tasks Take cuttings of your plant regularly to keep its shape bushy, and also as an insurance policy should your plant undergo a sudden decline.

Danger signs Leaves that go pale and rapidly drop off are usually a sign that there is too much water around the roots. Brown, dried leaf tips are usually blamed on dry air, but if your substrate is airy enough to allow generous watering, crispy tips should abate.

Toxicity No known toxicity for humans; toxic to pets.

Display Here's how my *B. maculata* grows: I have a large, flat-bottomed salad bowl, the base covered with a thin layer of leca. Sitting on top of that are several individually potted Begonias. I water from above directly onto the substrate in the pots, and any water that passes through the drainage holes ends up being absorbed by the pebbles. This is a great way of being able to mix and match different Begonia leaf shapes and colours; change the display weekly if you wish. This setup could easily be scaled down to accommodate just one *B. maculata* if you end up with a really large specimen.

Cultivars Once you have 'Wightii', start expanding your collection into the various other spotty cane begonia cultivars and hybrids. Silver-spotted 'Lucerna' turns into a meaty shrub and produces copious pink flowers while 'Looking Glass' is supremely silvery.

Also try . . . If you cannot keep cane begonias alive, try the beefsteak begonia (*Begonia* x *erythrophylla*). It was one of the first Begonia hybrids produced in 1845, and as it grows from a rhizome, it is super-tough.

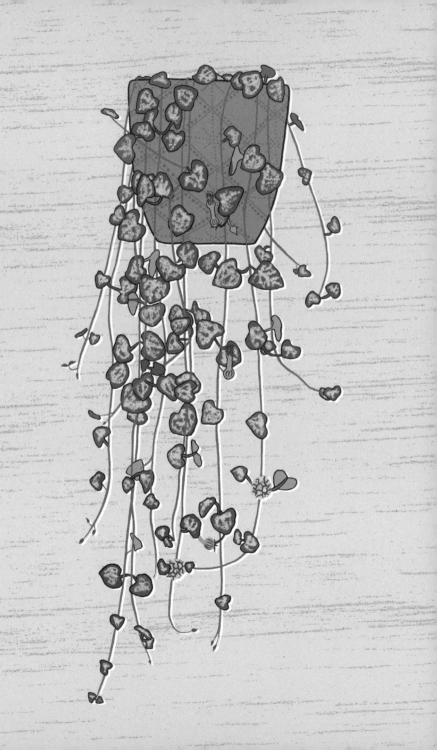

Ceropegia woodii

STRING OF HEARTS

Apocynaceae

G nats, no-see-ums, punkies, midges ... there are several common names for the 1,000 or so known fly species that make up the genus Forcipomyia, and yet these tiny creatures aren't much regarded by humans other than as a minor irritation. If you grow string of hearts, however, you may unwittingly find yourself at the centre of a Forcipomyia hostage situation.

Although it is the silvery, heart-shaped foliage of *Ceropegia woodii* that most often captivates the modern grower, the flowers warrant our attention too, not least because they function as temporary traps for the insects that pollinate them. When ecological scientist Jeff Ollerton and his colleagues studied the pollination practices of *C. woodii*, he discovered that this plant has an incredibly targeted approach to sexual reproduction. Its lantern-like lilac blooms only attract flies from the genus Forcipomyia, as opposed to the thousands of other species of small flies of a millimetre or two long that would just as easily fit inside. The plant growing in my conservatory will attract British-based Forcipomyia species, whereas string of hearts growing in a living room in Cape Town are pollinated by blood-sucking African species of Forcipomyia.

This tailored approach to pollination is seen across all the 200 or so species in the genus Ceropegia: they are, as far as scientists know, all pollinated by flies, but each concocts its very own chemical signal to attract specific fly species. Some Ceropegia flowers concoct volatile organic compounds that fool flies into thinking they are being drawn to sites where they like to lay their eggs. This strategy is called brood site mimicry; creating the smell of a rotting fruit, or dung, or other scents that seem delicious to flies with eggs to lay. Others mimic the scent of a fly's favourite food. Take the giant Ceropegia,

C. sandersonii, another popular houseplant species from southern Africa. Its white and green parachute-like flowers waft an array of volatile organic compounds that create the signature of an injured western honeybee to lure in its pollination partner, the carnivorous flies of the genus Desmometopa. These scents are not necessarily detectable by the human nose, although *C. cimiciodora* is so named because its perfume smells like bed bugs.

And *C. woodii*? As I write, we don't know how it pulls in its fly of choice. We do know that once the midges enter the flower's inner sanctum, they are held captive: that's why the German language uses the common name *Kesselfallenblümen* – kettle-trap flowers – for Ceropegia. The flower is shaped like a lilac-coloured turkey baster; at the base is the 'bulb', and at the top, an egg beater-shaped structure made up of a set of hairy, dark purple corolla lobes. The flies squeeze between the lobes and into the tube, which is both slippery and lined with downward pointing hairs that allow flies in, but prevent them from escaping. The hostages feed on the nectar that is secreted by the flower's corona at the base of the bulb. While there, the flies inadvertently pick up packets of pollen on their mouth parts, ready to deposit them on the next flower they visit. Within a day or so, the flower starts to wilt and the flies can finally exit. *Ceropegia woodii* grown indoors usually flower sporadically throughout the warmer months, although it's easy to miss the flowers: you may just notice a pile of withered turkey basters on the floor where the short-lived blooms have been jettisoned.

This plant grows wild in South Africa, Eswatini and Zimbabwe, but it has been popular as a houseplant since its introduction to the trade at the end of the nineteenth century. It is well adapted to cope with drought. The wiry stems emerge from a potato-like tuber up to 5cm in diameter that is rich in starch, storing water and nutrients to help the plant survive dry periods: although it is not a staple of the human diet, the tubers have been used as a famine food. Other Ceropegia species with more substantial tubers are regularly eaten, and in India (where there are around 50 endemic species of Ceropegia), traditional medicine recommends the tubers of *C. bulbosa* to treat stomach disorders and other conditions. Studies of Ceropegia tubers have found they contain the alkaloid cerpegin, and scientists are investigating whether this and other compounds in Ceropegia may have practical applications in medicine.

The tubers are usually found tucked between rocks where small soil pockets can support them; they sometimes populate sandy or rocky soils in forested areas too. The tubers are generally rooted in shade, while the vining stems scramble about in a mixture of sun and shade. As an insurance policy against being eaten, string of hearts endeavours to root itself wherever it can, producing aerial tubers along its stems – hence another of its names, the rosary plant, as the pea-sized tubers look like a string of rosary beads. These tubers not only make the plant hard to kill through neglect, they are amenable enough to be used by succulent breeders as rootstock for grafting trickier members of the related group of succulent plants called stapeliads, commonly known as carrion flowers. The spindly stems that emerge from string of hearts' tubers are irregularly clothed with pairs of heart-shaped leaves – purple on the underside, marbled silver and green on top. This silvery pattern isn't caused by a lack of chlorophyll pigment, but is a form of variegation known as blister or reflective variegation, as explained in the chapter on *Begonia maculata* (see page 21).

Ceropegia woodii has undergone a huge resurgence of interest in the last decade, but I am not sure it matches the level of hype around the species when it first came into cultivation in the US and the UK, if the price of plants is used as an indicator. In 1899, William Bull, a plant merchant on the King's Road, Chelsea, in west London, was selling *C. woodii* under the category 'new rare and desirable stove plants', for 10s 6d, which is about £70 in today's money. Two years later, Bull exported *C. woodii* to the US. This was just 19 years after the first reported sighting of the plant by a westerner. A report in *Curtis's Botanical Magazine* reports that British-born botanist John Medley Wood, then curator of Durban Botanic Gardens in South Africa, found a specimen 'hanging from rocks on Groen barg, Natal at 1800ft' – Natal being the name for the coastal South African province now called KwaZulu-Natal. The species was subsequently named after him. By 1904, English nursery H. Cannell and Sons in Swanley, Kent, advertised string of hearts for 1s 6d, around £9 today. Fast forward to 1908, and the Southern California Acclimatizing Association's catalogue listed the plant for just 25c.

At the turn of that century, string of hearts was designated a 'stove plant' ideal for a hanging basket, rather than as something to be grown indoors: unless you lived in California or somewhere simi-larly balmy, homes were far colder and the air more polluted than our modern, centrally heated interiors, so *C. woodii* would probably have struggled to thrive. The stove house – a glasshouse heated by a stove usually powered by coal – was therefore the more obvious place to raise string of hearts. The plant did move into homes at some point in the 1900s, however: by 1946, the *Rural New Yorker*'s 'Woman and Home' page lists *C. woodii* in an article on 'houseplants without direct sunlight'. The correspondent, referred to only as CW, describes a display on their living room table consist-ing of a miniature garden gate with trellis over the top and a pot of string of hearts on either side.

Ceropegia suffered a slump in the following decades. Garden writer Alan Titchmarsh in his 1982 book *The Hamlyn Guide to Houseplants* called it 'curious rather than beautiful'. Britain's foremost houseplant writer Dr David Hessayon was also rather dismissive in his bestselling *The Houseplant Expert*, the 1980 edition calling it 'an easy plant to grow' but warning that 'the foliage is unfortunately sparse and the 1in tubular flowers are insignificant'.

I like to think I played a very minor role in the species' revival in the last decade. In February 2013, Alys Fowler, gardening column-ist at the *Guardian*, wrote a column about houseplants for the newspaper's Saturday magazine. I was gardening editor and there-fore editor of her copy at the time. She called string of hearts 'the most forgiving of all houseplants', warning against overwatering, and suggesting it be grown in a bathroom. The only British nursery we could find selling this species at the time was by mail order, via Ralph Northcott's Cactus Shop nursery in Devon. The version of the article published online linked to their website, and within hours of the article's publication, they were sold out.

Since then, string of hearts has become a fixture of the houseplant scene, often reduced to the moniker 'SOH'. Perhaps the ultimate sign of the global acceptance of this plant is that IKEA now sells an artificial string of hearts.

CARE GUIDE

Light String of hearts can cope with a wide range of light levels, but benefits from some direct sunlight. Remember that in nature the tubers are in shade and the foliage in sunlight – albeit sometimes shaded by other vegetation. Light-starved plants will become extra-sparse and the leaves turn plain green.

Temperature It prefers a temperature range of 15-30°C (59-86°F) but will cope with short periods dropping down to 12°C (54°F) if kept dry.

Water The tubers will keep the plant supplied with water for periods when you forget to give it a drink; the greater risk is that the tubers and roots rot due to waterlogging of the substrate – see below. From spring to autumn, water when the substrate feels dry most of the way down the rootball; leave it to dry out completely between waterings in winter.

Humidity Not bothered by dry air.

Pests and diseases It seems fairly resistant to most pests, but woolly aphids, scale, spider mites and mealybugs can all try to take hold.

Substrate The more free-draining the better. Some growers even raise this plant in a saucer of gravel! Add grit, leca, perlite or another drainage material to houseplant potting mix to make up at least one third of the total volume.

Propagation When repotting, you may find the plant has produced several smaller tubers around the larger one. These can be teased away and potted up separately to make a new plant. If your plant produces aerial tubers, these can be looped back into the pot or into a separate container of substrate and pinned down where they will quickly root. Once established, cut away the stem connection to the parent plant. Slower, but equally effective, is simply cutting away a few stems and rooting them in a glass of water before potting up once the roots reach 3-5cm long. To make lots of new plants, remove lengths of stem and cut either side of each pair of leaves to create one-node cuttings which can be rooted on damp soil, perlite or sphagnum moss, covered with a clear plastic bag until established – this is sometimes called the butterfly method.

Feeding Feed occasionally with cactus and succulent fertiliser when in active growth.

Other maintenance tasks
Be prepared to sweep up the dead flowers regularly when the plant is blooming.

Danger signs If really left to dry out, the leaves will become soft and wrinkled, but do check the substrate before watering, as this can also be a sign of waterlogging.

Toxicity No known toxicity to humans or animals, although there are reports of contact dermatitis caused to greenhouse workers touching the foliage for extended periods.

Display Traditionally, string of hearts is displayed in a hanging basket or pot, or on a high shelf to allow the stems to trail extravagantly. Plants can look a little sparse; remedy this by propagating more stems and adding them into the pot. The stems can be trained around a wire trellis; they are not self-clinging, but the leaf shape means they hook onto each other and will stay in place with some support. If you leave them trailing, stems can grow as long as 2m or more: once they have reached the floor, simply chop off the excess and give away the cuttings or propagate new plants yourself. Terracotta pots will help to prevent waterlogging if you have a habit of overwatering; with the right substrate (see above) plastic works just as well. The stems can easily hook onto wandering pets, bringing pots down on furry heads, so keep them out of reach.

Cultivars Variegated cultivars have become popular, including the cream-splashed 'Variegata' and extra-silvery 'Silver Glory'.

Also try . . . Several other members of the Ceropegia genus are worth growing as houseplants; the parachute plant, *C. sandersonii*, has meatier stems and leaves and produces more substantial green and white flowers.

Chlorophytum comosum

SPIDER PLANT

Asparagaceae

In 2005, the BBC broadcast a short film called *Spider-Plant Man* as part of its annual Comic Relief charity fundraiser. The eponymous hero was played by Rowan Atkinson, who is best known for his comedy character Mr Bean. In a parody of the Spider-Man films, Atkinson played nerdy photographer Peter Piper who gets bitten by a genetically modified spider plant during a visit to Very Safe Labs Incorporated, thus transforming into the weediest (excuse the pun), green Lycra-clad superhero you can imagine.

Although this hastily put together film was firmly tongue in cheek, the BBC inadvertently tapped into something important about this houseplant species. Like Atkinson's character Mr Bean, the spider plant is instantly recognisable around the globe, has been ascribed various superpowers across its history, and is extremely good at world domination, with perhaps only devil's ivy (*Epipremnum aureum*) and *Aloe vera* challenging it for the title of the world's most widespread houseplant. The spider plant is underrated and unsung, the living wallpaper to so many of our lives. It makes such a useful houseplant because it will tolerate considerable neglect, and endlessly reproduces itself in a way that encourages its owners to spread its offspring far and wide.

There are about 230 species in the Chlorophytum genus, and *Chlorophytum comosum* is by far the most famous, although Indian species *C. borivilianum*, known as *safed musli* in Hindi, is a valued herb in traditional medicine, earning it the title of 'nature's Viagra', even though its effects have not been fully explored by scientists. *Chlorophytum comosum*, however, has a wide indigenous range across tropical Africa south of the Sahara, particularly on the eastern side of the continent, as well as Ivory Coast, Liberia and Sierra

Leone in the west. It's not quite accurate to call it a species, either: botanists prefer to call it a species complex; in other words, they are still not exactly sure where one species begins and another ends.

Scientists have distinguished three different varieties in the wild through DNA analysis: the slim-leaved *C. comosum* var. *comosum*, to which our domesticated spider plant seems most allied; plus *C. comosum* var. *bipindense* and *C. comosum* var. *sparsifolium*, which have wider leaves. All three of these grow in a similar fashion in the ground and can be found in a variety of habitats, including evergreen rainforests and on riverbanks, on cliffs and rocky slopes, in a range of soils, most of them slightly acidic. It occasionally experiences light frosts, but can also cope with temperatures of 30°C (86°F) or more, while the fleshy white roots allow the plant to survive in rocky clefts as well as moist soil. This explains why it makes a successful indoor plant, able to make the most of the confines of a nursery pot and store supplies for occasions when the watering can is forgotten.

This species is poor at producing seed: instead, it reproduces vegetatively, using a clever adaptation of whippy stems known as inflorescences or runners. Each can reach 60cm or more, and some plants in cultivation will produce dozens or even hundreds of them. Each runner may start out growing one or more small white six-petalled flowers, which produce seed pods if pollinated, but often the flower is replaced by a baby plant. These plantlets – sometimes whimsically known as spiderettes – develop at scale-like nodes along the stem, and are an example of a phenomenon called pseudovivipary. In nature, these grow until they weigh down the stem enough to touch solid ground, and then begin to root. The Afrikaans name for the plant, *hen-en-kuikens* – 'hen and chicks' – reflects the way the mother plant sits surrounded by its babies, covering the ground in large thickets. It's up to you whether you leave the plantlets attached, or remove them.

The spider plant's history as a houseplant goes back more than two centuries. Swedish botanist Carl Peter Thunberg described it in his 1794 work on plants he had seen and collected in South Africa, *Prodromus Plantarum Capensium*. He called it *Antheri-cum comosum*, but it was bounced around several genera during the 1800s, including Phalangium, Caesia and Hartwegia, before being settled as Chlorophytum in 1862. *Chlorophytum* simply means 'green plant', while *comosum* derives from the Latin word *coma* meaning tuft or hair, referring either to the overall look of the plant, or of the plantlets. It was not until the 1880s that variegated spider plants began to be available, gradually overtaking the green-leaved specimens in popularity.

It is interesting to speculate whether the spider plant's fecundity has also led to its reputation in traditional medicine as a charm for pregnant mothers and newborn babies. It is known as *ujejane* by the Xhosa people and *iphamba* in Zulu, and has a history of use around pregnant women and babies: both through a decoction of the roots to newborns as a purgative, and as a protective charm for mother and child. In Chinese traditional medicine, *C. comosum* is used to treat bronchitis as well as coughs, fractures and burns. Western medicine has also become interested in this plant: it is packed full of chemical compounds called steroidal saponins, and scientists are beginning to explore their possible applications in cancer treatments.

Meanwhile, modern witchery seems to have embraced the spider plant as a fertility symbol. Those wishing to plan a pregnancy are advised to hang a spider plant in the bedroom, while witch and herbalist Juliet Diaz suggests in her book *Plant Witchery* that keeping a spider plant in every room will harness the plant's role as 'energetic filters, kind of like dream catchers for negativity'. The plant seems to garner a rather different reputation in South America, however, where it is known as *malamadre* ('bad mother'), presumably for its habit of producing more offspring than any actual mother could handle.

The common name of spider plant is often thought to refer to the plant's spidery silhouette: the appellations of the airplane plant and the ribbon plant probably have a similarly simple explanation. There is another, albeit convoluted, explanation for the spider reference, however. Thunberg's original classification of the species as *Anthericum comosum* placed it in the same genus as a similar-looking southern European plant called spiderwort, *A. liliago*. This plant was, since ancient times, thought to have powers as an antidote to spider bites. Both were later reclassified under the genus Phalangium, which is also the name of a genus of harvestmen (spider-like creatures that are classed as arachnids, just to add another layer of complexity). Ultimately, our familiar spider plant ended up in

Chlorophytum, and the European plant moved back to Anthericum, but the linguistic link between the two remained. Interestingly, the spider plant also stole one of *A. liliago*'s common names, St Bernard's lily, and you may see *C. comosum* sold under this name today. Older houseplant books also refer to the plant as walking anthericum.

Another common name you may come across is Goethe's plant. That is because the spider plant's most famous fan was Johann Wolfgang von Goethe, the legendary German poet with a passion for botany. He was given a specimen by the Grand Duke of Saxe-Weimar, Carl August, who had bought the plant in Leipzig in 1827, but neither man knew what it was. Goethe quickly became fascinated, writing to a friend and marvelling at its 'extraordinarily prolific habit which unfolds before our view the whole life of a plant'. Like all good spider plant owners, he sent specimens to his botanist friends. The name Goethe's plant is also applied to an equally prolific houseplant of which the poet was fond, the Madagascan succulent *Kalanchoe pinnata*, which grows plantlets along its leaf margins.

In the 1980s, the spider plant gained another alleged superpower – cleaning the air of damaging air pollutants, specifically formaldehyde, which is often found in soft furnishings and furniture around the home. A scientific study carried out by Dr Bill Wolverton on behalf of NASA tested spider plants alongside two houseplant species – golden pothos (*Epipremnum aureum*) and arrowhead plant (*Syngonium podophyllum*) – to see which would remove formaldehyde most efficiently from the air in a sealed container. *Chlorophytum comosum* came out on top, and since then a whole host of experiments have tested its air-purifying powers. While it is true that spider plants do remove pollutants from the air around them, the sticking point has always been the number of plants that would be required to create a significant difference to air quality. Dr Wolverton's experiments, including his better-known and larger study published in 1996, tested plants in sealed environments rather than the unsealed territory of an average home. Even if you placed a *C. comosum* in every room, the impact it would have on your air would be negligible: estimates of how many specimens you would need per square metre vary, but are high enough to render the idea pointless for the average home. Nevertheless, thousands of articles on houseplants still over-egg Wolverton's research, suggesting that a spider plant on your desk is all you need for clean air.

This is, of course, not a reason to reject your spider plant: this survivor continues to be one of the hardest houseplants to kill, so the care guide that follows may seem superfluous. Just because a spider plant can withstand huge levels of neglect does not mean it should. Grow a *C. comosum* well and you will be rewarded with one of the houseplant world's most delightful displays: a chandelier-like mass of leaves, runners and plantlets, just waiting for your next guest to pass on its genetic heritage to another household.

There is a downside to the spider plant's popularity. It has become an invasive species in countries where the climate is warm enough that abandoned plants quickly take root, particularly Australia. If not controlled, the spread of this and other non-indigenous plants damages the biodiversity of sensitive ecosystems and leads to the demise of indigenous plant species.

CARE GUIDE

Light Spider plants will adapt to a wide range of light conditions; too much sun and they will begin to look bleached, but they can cope with morning sun, especially in winter. In deep shade, variegation will be less pronounced and the plant will idle in neutral.

Temperature Temperatures of down to 7°C (45°F) or even lower can be tolerated, but plants will be happier at a room temperature of 18–23°C (64–73°F).

Water The meaty white roots can hold enough water to enable spider plants to live for many weeks without water. But plants look best when they are watered more regularly. Use rainwater if you can, as water containing fluoride can lead to blackened leaf tips.

Humidity It's not essential to boost humidity for this plant, although blasts of hot, dry air from heating systems will weaken its constitution.

Pests and diseases Not particularly prone to pests, although thrips, mealybugs and red spider mites can occur.

Substrate The roots will take over the whole of a pot when growing happily, so do check the rootball regularly and repot when necessary. Any regular houseplant potting mix will do.

Spider plants grow well in hydroponic systems.

Propagation Once plantlets have reached a decent size, pin one onto the surface of a small pot of houseplant potting mix while leaving it attached to the runner to allow it to root with minimal stress. Alternatively, snip plantlets off the runner and root in a glass of water; or where plantlets have already developed roots, pot them up individually.

Feeding Apply a foliage houseplant fertiliser regularly when the plant is actively growing.

Other maintenance tasks
Placing the plant high up to enjoy its cascading habit can mean problems go unnoticed. Regularly take your plant down off its perch to check for pests.

Danger signs Some plants will be slow to produce runners. This is usually due to a light deficit: moving plants to a brighter spot should help. Bleached out leaves signal the plant is being burned by the sun.

Toxicity No known toxicity.

Display Some growers allow the runners and plantlets to remain, giving a chandelier effect which looks best displayed so it can be viewed on all sides, on a plant stand or in a hanging pot. There was a trend in the 1980s for training the runners up a hoop or wire, but this seems to have fallen out of favour. Spider plants can also be used as ground cover in larger indoor landscaping displays.

Cultivars Two main variegated varieties used to reign supreme: 'Vittatum', with green-margined leaves either side of a central cream stripe, and creamy white runners, and 'Variegatum', with a green central stripe and creamy margins, as well as green rather than yellowish runners. There are several cultivars available, including 'Bonnie' with its twisted leaves, and the lime-green 'Lemon'. The plain green *C. comosum* is a relative rarity in cultivation.

Also try . . . If you like tough houseplants with long strappy leaves, the epiphytic cactus *Pfeiffera boliviana* is an excellent choice.

Crassula ovata

JADE PLANT

Crassulaceae

Deep within a mountain in Hebei province, central China, the tombs of Liu Sheng and his wife Dou Wan lay undiscovered for more than 2,000 years. It was only when the People's Liberation Army of China set to work blowing chunks out of the slopes of Lingshan Mountain in 1968 during construction work that they chanced upon an entrance to the burial chamber of this king of the Han empire, who died in 113 BCE. Chinese archaeologists called in to investigate found both bodies were encased head to toe in a carapace made of thousands of jade tiles held together by a network of gold wires. Jade burial suits had, until then, been assumed to be mythical, existing only as accounts in ancient books. Here was the first proof they were real.

Jade – *yù* in Mandarin – is a collective name for the minerals nephrite and jadeite. It is highly prized in Chinese culture as a symbol of immortality, purity and prosperity. Although the physical bodies of Liu Sheng and Dou Wan rotted away, they did achieve a form of immortality through their elaborate preparations for the afterlife. And the market for jade remains buoyant, as the expanding Chinese middle class invests in jewellery and other objects celebrating this symbolic substance.

No wonder, then, that a plant whose fat and glossy green leaves resemble chunks of jade carries a heavy freight of meaning in China. As a symbol of longevity, *Crassula ovata* has excellent credentials, as the cuttings sitting on the shelf in my office illustrate. I cut these chunks from the parent plant many weeks ago, and yet they are just as shiny and green now, and almost as succulent. The cut ends are just starting to push out roots; as soon as I place them in soil they will begin to grow, none the worse for their hiatus on a shelf.

45

Reverence for *C. ovata* extends across the globe, although outside China it is valued more for its symbolism as a bringer of prosperity than immortality. Visit Chinese takeaways and restaurants from Bedford to Buenos Aires and you will find one by the door – a location that, according to the Chinese philosophy of feng shui, will invite wealth inside. American feng shui practitioners even suggest putting a 100 dollar bill under the pot of your jade plant so it can 'grow' cash. In reality, the only way a plant will make you money is if you sell it: large specimens sometimes retail for hundreds of pounds.

Despite its deep associations with China, *C. ovata* comes from Africa. It grows wild in the arid or semi-arid Eastern Cape and KwaZulu-Natal provinces of South Africa, and is most populous on the coastline, occupying rocky slopes, forests established on ancient sand dunes, and dry river valleys. Temperatures in these regions can reach 40°C (104°F); rainfall is low.

The species was one of the first succulents that became popular in cultivation in Europe, as Africa was colonised by the West in the seventeenth and eighteenth centuries; the exact date of its arrival in Europe varies according to which account you read, but could date back as far as 1690. It was given the species name *Cotyledon ovata* by Scottish botanist Philip Miller in 1768, who was superintendent at the Chelsea Physic Garden. John Loudon's book *The Green-house Companion* of 1828 described the species as 'of little beauty'. It wasn't until 1916 that it joined the genus Crassula, thanks to the taxonomic work of botanist George Claridge Druce. (Although he was a pharmacist by training, his real passion lay in botany, about which he wrote many books.) The jade plant is still sometimes known by another outdated scientific name, *C. argentea*.

Plants can grow into multi-stemmed shrubs up to 5m tall, with grey, gnarled trunks up to 20cm thick: chunky enough to play host to wasps' nests on occasion. However, they don't always reach their full height in the wild, because the juicy leaves are grazed by all sorts of creatures: taller animals such as giraffe, rhino, springbok and kudu browse the tops and sides, while leopard tortoises nibble from below. Humans have made use of this plant too, ascribing to it a number of applications in traditional medicine; in India, people drink juice from the leaves to treat diabetes, while European settlers and indigenous Bantu-speaking people in Africa applied a cut leaf to warts. The grated roots were also cooked and served with milk to treat diarrhoea.

Other succulents have evolved strategies to render themselves unattractive and unpalatable to herbivores by sporting a thick coating of hairs or a pincushion of spines, or filling their veins with toxic sap. For the jade plant, though, becoming a snack is a chance to spread. A section of stem, a single leaf or even a part of a leaf that drops to the ground as herbivores munch messily will quickly develop roots and start a new plant. The branches are also brittle enough to break off in high winds or even under the pressure of their own weight as the plant grows large, creating more opportunities for propagation.

Just like my cuttings, these discarded chunks manage to avoid drying out completely using the same adaptation that the parent plant employs during droughts. CAM may sound like motor racing jargon, but it refers to crassulacean acid metabolism (CAM). This specialised form of photosynthesis allows plants to cope with arid conditions by opening its stomata (breathing pores) at night to take in carbon dioxide, and closing them by day when temperatures are at their highest, in a reversal of the usual regime for non-CAM plants. They convert the gas into malic acid and store it, then come daylight they convert it back to carbon dioxide so photosynthesis can occur. When times are really tough, another process known as CAM idling kicks in: stomata remain shut night and day, while the plant recycles carbon dioxide already in its tissues.

The genus Crassula has its root in the Latin word *crassus*, meaning fat, while *ovata* means egg-shaped, although botanists will tell you that, strictly speaking, this species has obovate leaves: teardrop shaped, with the tapered end joining the stem. Different cultures have found other ways of describing the foliage: one of my favourite common names is *plakkies*, which means flip flops in Afrikaans, while I have heard that in Shanghai it is called ping pong bat plant in Shanghainese. I have seen many references to jade plants being called cauliflower ears in China too, but have not managed to find a logical explanation of why that name applies: again, perhaps the shape of the foliage is responsible.

The leaves contain a further mystery. They are encased in a waxy, protective coating known as a cuticle, which helps to slow down water loss. Amid the glossy sheen of this cuticle, a close examination reveals a smattering of tiny white dots that sometimes get mistaken for some sort of disease or pest. Each of these dots marks the location of a hydathode: a pore that is roughly double the size of the plant's stomata. The dots are made up of mineral deposits that come from water. Hydathodes are usually associated with secretion of water, known as guttation, a phenomenon that allows plants to expel excess water in moist climates. Given the jade plant's arid habitat, botanists have been baffled that they are a feature of this and several other Crassula species that originate from the same region. The favoured theory is that hydathodes are used by these succulent plants to reverse the guttation process and collect water in the form of night-time fogs and dews, which would make sense given the plant's coastal distribution and the lack of rainfall.

What of the flowers? They are not the species' best feature, but at least one can say they do not smell like cheesy feet, as some smaller houseplants in the Crassula genus do, such as *Crassula muscosa*, the watch-chain succulent. In the wild, plants flower during the cooler months of June, July and August, putting out masses of starry flowers that are mildly soapy in scent. They may not produce flowers in their first few years; often specimens will have reached 1m tall before blooming. Flowering is triggered by short day length, so will usually happen in December or January in the northern hemisphere. If your plant fails to flower, make sure it is not exposed to artificial light after dark. Feng shui adherents, though, may wish to stave off flowering, as this is associated with mourning and funerals. The flowers are pollinated by flies, bees, butterflies and beetles, and the seeds that follow are tiny enough to be dispersed by the wind.

CARE GUIDE

Light A few hours of direct sunlight a day is ideal. Plants that do not get enough light will start to become leggy, and new leaves may become smaller.

Temperature A minimum temperature of around 10°C (50°F) can be tolerated in winter provided the plant is kept dry.

Water Ease back watering from autumn to early spring: if your plant is in a heated room, only water when the leaves become a little soft and wrinkled. During late spring and summer, water once the substrate is dry at root level.

Humidity They are adapted to life with low humidity, so there's no need to make the air moist.

Pests and diseases Mealybugs.

Substrate An over-the-counter cactus and succulent mix will suffice, or combine two thirds houseplant potting mix or John Innes No. 2 with one third grit or perlite.

Propagation Stem or leaf cuttings will quickly root; once cut, allow the wound to dry out for a couple of days before placing the cut end in gritty potting mix.

Feeding Use cactus and succulent fertiliser regularly when the plant is in active growth in spring and summer.

Other maintenance tasks Plants that get leggy or out of shape can be pruned back in spring.

Danger signs Yellowing leaves are usually a sign of waterlogging at root level. Sudden leaf drop can also occur in this scenario. Lower leaves will occasionally shrivel and drop.

Toxicity No known toxicity.

Display A terracotta or glazed china pot is a good choice once plants get large, preventing them from becoming top heavy. Jade plants make a good subject for indoor bonsai, where the roots and stems are pruned carefully to reduce the size of the plant: in this case, choose a shallow bonsai-style pot.

Cultivars There are a number of cultivars of *C. ovata* on the market: 'Gollum' and 'Hobbit' have leaves that look like green coral, while 'Variegata' has green- and cream-streaked leaves. 'Crosby's Compact' may be particularly useful for those wanting to try a jade plant bonsai. My favourite, though, is 'Hummel's Sunset', whose leaf margins turn bright red and yellow in a sunny summer.

Also try . . . If you lack space, try the more compact and smaller-leaved *Portulacaria afra*, known as the miniature jade.

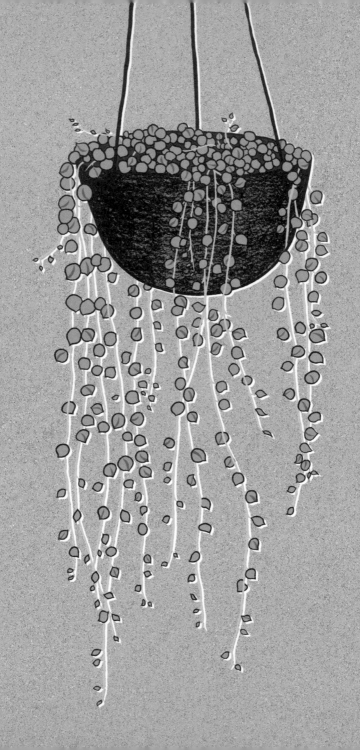

Curio rowleyanus

STRING OF PEARLS

Asteraceae

W ere the early twenty-first century boom in houseplants to be given a Hollywood film treatment, I like to think the movie would be called – cue portentous voice – *Houseplants: The Rise of the Strings*. The soaring popularity of wiry-stemmed, trailing species that look good tumbling down a shelf or suspended from a macramé hanger has been one of the defining trends of the last few years. String of hearts, string of spades, string of lemons, dolphins, bananas or buttons; take an object of your choice that could conceivably echo a leaf shape, add the prefix *string*, and there's a good chance you just named a houseplant.

The champion of them all, though, is string of pearls, *Curio rowleyanus*. One of my proudest houseplant achievements was growing a pot of this plant that poured over a high shelf like a green waterfall more than 1m long. And yet this plant is not grown as intended by its evolutionary path. Despite its popularity as a trailing plant, this species is adapted to horizontal life: in nature, it scrambles about, forming a dense carpet over rocky ground that resembles a beaded car seat cover. Admittedly, 'mat of pearls' doesn't have the same ring to it, though.

Then again, this plant is far more prolific as a houseplant than a wild plant. Compared with many of the species in this book, string of pearls is a new introduction into the houseplant world. The millions of plants in circulation now can be sourced back to one specimen that was sent to British grower John Measures of Keyston, Northampton, in 1950. How did Measures get hold of it? Information is scant. The species was given the scientific name *Senecio rowleyanus* by German succulent expert Hermann Jacobsen; this information was published for the first time in the journal of the National Cactus and Succulent Society (a precursor to the current British Cactus and Succulent Society, the BCSS) in 1968; string of beads was the first English common name it was given. 'Unfortunately all efforts to trace its place of origin have failed,' Jacobsen wrote. All we know is that the plant Measures obtained came to him via an intermediary who reported that a collector had found it in south-west Africa. The species was subsequently located in the wild a few years later in the Eastern Cape province of South Africa, but it is by no means as widespread as other related species.

Jacobsen named the plant after his friend and fellow succulent expert, botanist Gordon Rowley, a prolific author and president of the BCSS. He died in 2019, two years after his final book was published, at the age of 95. We have him to thank for popularising and defining 'fat-bottomed' succulents as caudiciform plants; he first coined the term in 1948. In 1997, taxonomic research concluded string of pearls should be removed from the genus Senecio – often called a 'mega genus' because it is so large, with 2–3,000 species – to the new and far smaller genus Curio. 'Curio' is usually defined as a small unusual object that people collect, which is an apt name for this species, but English botanist Paul V. Heath chose this name as a reference to the Roman statesman and orator Gaius Scribonius Curio: or rather orators, as there was a father and son with the same name who lived in the last century BCE.

What do we know of how *C. rowleyanus* grows in the wild? Jacobsen described it as 'a low-growing herb with prostrate slender stems rooting adventitiously to form dense mats'. If we refer to its close relative, *C. radicans* or string of bananas, it is not unreasonable to presume its territory may be very similar. *Curio radicans* is a ground cover plant which is widespread in the Cape provinces of South Africa. It has a fibrous, shallow root system that takes every opportunity to put out adventitious roots along its stems to anchor itself in the ground; this also provides insurance against being wiped out if herbivores browse on parts of the plant. It grows in rocky places, often proliferating under rock ledges or bushes where it gains a degree of protection from the sun.

This tells us a great deal about how *not* to grow string of pearls as a houseplant. Of all the poorly succulents I have seen, *C. rowleyanus* is one of the most frequently abused. Large scale nursery production dictates that plants are sold in plastic hanging containers with inadequate drainage holes, containing a medium that holds too much moisture for the roots to be happy in the conditions of the average home. That is a lethal combination for this species, because the roots of string of pearls are fine, shallow and an easy target for rot if exposed to too much moisture and not enough air.

Some succulents store water in their stems, like the caudiciform plants championed by Rowley, but string of pearls stores water in its spherical leaves. Well, almost spherical; each 'pearl' is tipped with a mucro, a botanical word for the pimple-like protrusion opposite the point where the leaf joins the stem. This shape maximises water storage potential in the inner tissues of the plant, while reducing the area of the outer surface of the leaf, through which water is lost. Look closely at a leaf, particularly when it is backlit by the sun, and you should see a darker, translucent stripe running across it. This leaf 'window' allows the plant to carefully control the amount of light passing into its interior; vital for managing temperatures inside the tissues and preventing damage from the sun's radiation.

It is far easier to spot the leaf windows in the variegated form of *C. rowleyanus*, which has cream-striped leaves, but the windows remain green. The leaves remind me of mint humbugs, although Rowley described the variegation as resembling 'peas with mayonnaise'. Like the species, the origins of this variegated form also remain a mystery, other than vague indications that it originated in the US. Rowley gave it the cultivar name 'String-of-Pearls', and strangely, this common name has crossed over to the plain green species as well. In a clever piece of marketing, American mail order firm Crane Norris of Freeport, New York, ran press advertisements in the 1970s claiming the plant as some sort of religious miracle, renaming it the rosary bead plant and selling it via mail order for $5 a go. This common name persists today.

Although string of pearls is primarily considered a foliage plant, flowers do appear once plants are mature. Small clusters of white flowers with purple or red stamens tipped by yellow anthers appear in winter, wishfully described by some as a disco ball, but resembling a tiny firework in my opinion. Your nose may become aware of them before your eyes do; the scent is often described as cinnamon-like, but some perceive a carnation-like clove smell. The seeds that follow are dispersed by the wind, each sporting a pappus – a mass of hairs rather like a dandelion seed – that allow it to become airborne with ease.

The scent of the flowers may be mouth-watering, but there is no record of any part of this plant being eaten, or of any medicinal applications. Fears did circulate within the American succulent world that these plants were deadly poisonous. Gordon Rowley reports in his book *Succulent Compositae* that 'eventually this malicious gossip so incensed one courageous lady at a society meeting that, in front of a gasping audience, she seized a plant, tore off a piece and ate it just to prove that it was harmless!'.

You may notice that I talked about my own specimen of *C. rowleyanus* in the past tense. This isn't strictly correct: I still have the plant, but it is a shadow of its former self, and I can no longer boast about my monster string of pearls. I did not pay enough attention to it one winter, and thought the shrivelled leaves indicated that the plant needed more water. I was wrong: it was rotting. I wonder now whether the weight of the stems, which were more than 1m long and fairly heavy, put too much strain on the plant and this, combined with too much water, caused the root system to collapse. After all, in nature the weight of the stems would be supported rather than hanging free, and the root system would be spread across a wide area of the plant rather than concentrated in one spot. I managed to salvage and root a few healthy strings. Now, I trim any stems that get too long, plant them in a shallow, clear plastic container so I can keep an eye on the roots, and water with extreme caution during winter.

CARE GUIDE

Light This plant does not grow in full sun in the wild, but it is likely that light levels in your sunniest room are still lower than in its habitat in the Eastern Cape. A conservatory, south-facing window or large west-facing window is ideal.

Temperature It can tolerate a minimum temperature of around 10°C (50°F) in winter provided it's kept dry.

Water Provided the substrate and drainage are right (see below), as with all succulents, watering fairly generously in summer is not a problem. Come winter, though, it's a different story. Allow the plant to dry out during autumn, and only water enough to stop the leaves from shrinking dramatically before spring arrives.

Humidity String of pearls laughs at dry air. A combination of moist air, cool temperatures and too much water at the roots is a recipe for root rot.

Pests and diseases Mealybugs.

Substrate If your plant is still in the substrate you bought it in, repot ASAP. A well-drained substrate is necessary to avoid root rot: you could risk an over-the-counter cactus and succulent mix, but if you can, create your own substrate by combining two thirds to one half houseplant potting mix or John Innes No. 2 with one third drainage material, such as perlite or fine grit.

Propagation Even a single leaf will grow to a new plant eventually, but it is much quicker to take stem cuttings. Snip off a few stems around 5–10cm long with clean scissors, leave to dry for a day or so, then gently press into the surface of a tray of gritty potting mix. They should produce roots within a few weeks; mist the cuttings to keep them moist in the meantime. Alternatively, you can strip the leaves from the lower centimetre or two at the base of the cutting, and insert this bare end into the substrate to root. Or just loop some of the longer stems back onto the surface of the pot, and weigh down with a few small pebbles or coins. They will root easily: this is a great way to create a fuller pot. I have also had success rooting cuttings in a glass of water and even in a damp, clear plastic bag, so it's worth experimenting if one method fails.

Feeding Use cactus and succulent fertiliser regularly when the plant is in active growth in spring and summer.

Other maintenance tasks Provided your setup is on point, this plant should not need a great deal of care.

Danger signs Watch your plant like a hawk: the pea-like leaves will show when conditions are not right. They should be firm and filled out – signs of leaves dissolving, turning to mush or shrivelling are indications that something is wrong at root level. In winter, the leaves may become slightly soft and shrivelled when the substrate is really dry, but water only enough to reinflate them.

Toxicity Mildly toxic to pets; no known toxicity to humans.

Display Hanging pots or a position on a high shelf or in a wall pocket are ideal, but ensure pots have plenty of drainage, and take them down to water so there is no chance of them sitting in a stagnant pool. Given the plant's shallow root system, you could opt for a 'mat of beads' look and display a plant in a shallow pan, allowing the stems to snake around and root into the substrate. Once stems become very long – over 1 m, say – snip off to propagate or offer to friends. I prefer a clear plastic pot so I can observe what is happening to the roots, but if you tend to water too much, opt for porous terracotta – shallow, rather than deep.

Cultivars The only cultivar I have come across is Rowley's variegated 'String-of-Pearls'.

Also try . . . Why not create your own curtain of succulent strings? *Curio herreanus* is similar although its leaves resemble gooseberries – it is also known as string of raindrops – while string of bananas, *C. radicans*, is another excellent choice. *Crassothonna capensis* has purple tinged stems and leaves shaped like spindly beads, earning it the common name ruby necklace.

Dieffenbachia seguine

LEOPARD LILY

Araceae

If you have only read breezy blurbs for the lush tropical delights of Dieffenbachia on houseplant websites, prepare yourself. This species has a dark backstory involving torture and domestic abuse, Nazis and slave owners.

The first thing you need to know about the leopard lily is its toxicity. At different points in history, humans have harnessed Dieffenbachia's powers to harm others. Most aroids and many other species grown as houseplants contain calcium oxalate in crystal form; it is also present in many foods we enjoy, including beetroots, spinach and edible members of the aroid family including taro and tannia. Dieffenbachia contains three different types of calcium oxalate crystals, and their distribution varies depending on the type of tissue concerned: stem, leaf, root and so on. The crystals that seem to do the damage are the needle-like raphides, which are held in bundles within cells shaped like spindles, with nipple-shaped ends; these are called biforines. Around the raphides sits a mix of proteolytic enzymes and other chemicals, so when the biforines are damaged by munching herbivores, the raphides and the chemicals are fired out, piercing the skin: what biochemist Gary G. Coté refers to as 'microscopic poisoned darts'.

Scientists are still considering the exact chemical cocktail, and the way the raphides do their damage, but we know the symptoms that follow very well. Assuming you are not put off by the warning smell of musk *Dieffenbachia seguine* releases when bruised, any contact between your mouth and the sap of this plant results in burning, excessive salivation and dramatic swelling of the tongue, throat and mouth parts. The common name dumb cane refers to the fact that those afflicted will be unable to speak. Bare skin that comes into contact with the sap may develop a red, burning rash and blistering. Sap splashed or rubbed into the eyes results in swelling, burning, watering and involuntary twitching. (No wonder Dieffenbachia is known as *ağlayan çiçek* – crying flower – in Turkey.) In extreme cases, a dose of Dieffenbachia can result in vomiting, diarrhoea and a swelling of the throat that can necessitate a tracheotomy to provide an opening into the windpipe. Thankfully, Dieffenbachia's acrid taste means most people do not swallow much, preventing most victims from ingesting a fatal dose, and after a pretty unpleasant couple of days, most people recover. Statistics on accidental plant poisonings in the US from 1985 to 1994 show that Dieffenbachia was not the plant that most frequently poisoned Americans: it is second to Philodendron, with Euphorbia in third place. There are periodic panics about 'deadly Dieffenbachia' in the popular press, and the internet has spawned a number of urban legends that this plant can kill you if you dare to simply touch a leaf, but unless you bruise or damage the plant, you will suffer no ill effect.

Dieffenbachia's ability to temporarily silence its victims was exploited in the most despicable way by slave owners in eighteenth-century Caribbean plantations who rubbed cut Dieffenbachia stems in the mouths of 'unruly' enslaved people as a punishment. There are also reports that enslaved people tried to commit suicide by eating Dieffenbachia; when mature, the stems resemble sugarcanes, so there are records of poisoning via mistaken identity too. This is one tiny sliver of the mass cruelty of the colonisation of the Americas, and just one aspect of the colonial botany that saw plants from the New World as 'green gold' to be explored and exploited. Perhaps it seems more shocking because the violence was done using a plant that we are used to seeing in our homes.

As we begin to face up to this ugly past – and realise how its legacy continues to shape our present – I believe the common name dumb cane should be consigned to the past, along with another name for *D. seguine*, the mother-in-law plant. Some find it tempting to write this off as a joke, but scratch beneath the surface of the silliness and a rich vein of misogyny centred on the silencing of women is revealed. Take this example: in 1916, several American newspapers carried the story – possibly an urban legend? – of John Kearne, who, after discovering the poisonous powers of Dieffenbachia on a visit to Shaw's Botanical Gardens in St Louis, Missouri, tried it out on his wife to stop her talking. His wife – she is invariably not named – ended up in hospital after unwittingly sampling the plant. Every account treats the story as an amusing tale illustrating Kearne's ingenuity, not as an assault.

In the wild, Dieffenbachia's toxicity is not aimed at humans of course, but at hungry herbivores. This does not stop all creatures from interacting with the plant. If you see Dieffenbachia in the wild, it is worth peering into the leaf axils – the point where the leaf joins the main stem – to see if there are any tiny tadpoles swimming in the water that pools there. Several species of poison dart frogs use these tiny ponds, known as phytotelmata, to raise their tadpoles, including the mimic poison frog, *Ranitomeya imitator*, with its black and yellow-green giraffe-like patterning. Once the eggs have hatched, the male frogs transport tadpoles on their backs and deposit them in carefully selected pools, where the young feed on detritus, algae and small creatures such as mosquito larvae.

There are roughly 50 species of Dieffenbachia, their home range stretching from Central America to Brazil; *D. seguine* is found in parts of the Caribbean and South America, including Colombia, Peru and Bolivia. Dieffenbachia is terrestrial: usually found in brighter clearings in humid or wet tropical forests, on riverbanks or swampy ground. When potted, plants can reach a great height, but in the wild, stems taller than 1 m tend to topple over and grow horizontally. Roots form along their length, and where the stem is damaged by animal tracks, it branches and shoots up again, slowly increasing one plant's real estate, so large clumps are not unusual. You may have noted that I couched mention of Dieffenbachia species with the word 'roughly' – that is because the genus has not received as much attention as other aroids, which have historically been understudied anyway. Aroid scientist Thomas Croat suggests that this is because cutting and preparing specimen plants carries the risk of injury.

Dieffenbachia was one of the earliest genera to be brought to Europe for cultivation, in 1759. It was moved around from Arum to Caladium before being settled in the genus Dieffenbachia when it was described by Austrian botanist Heinrich Wilhelm Schott in 1829. He named the genus after Joseph Dieffenbach, head gardener of the Imperial Gardens at Schönbrunn Palace in Vienna. The genetic heritage of the plants in cultivation today is largely lost, but the majority of the plants we grow are believed to have their origins in *D. seguine*. Several species including *D. picta* have been subsumed into *D. seguine* as research begins to clarify Dieffenbachia's family tree.

Like its fellow houseplant aroids including Spathiphyllum and Monstera (see pages 177 and 123), the 'flower' consists of a hood-shaped spathe – an adapted leaf – and a spike-like spadix covered in tiny individual flowers: the female flowers at the bottom of the spadix and the male flowers at the top. The spadix is not as prominent as in those species, generally being somewhat hidden by the foliage, and the spathe is green rather than white. Dieffenbachia's pollinators are beetles; often dynastine scarab beetles (if no reader decides to take Dynastine Scarab Beetles as the name of their up-and-coming rock band, I shall be disappointed). At dusk as the plant prepares to bloom, the beetles are drawn by the scent of the flowers to feed upon the starch-rich, club-shaped staminodes: infertile stamens that surround the female flowers. They crawl down into a chamber formed by the lower part of the spadix and enclosing hood of the spathe. After two or three nights of feasting, the beetles climb the spathe and are covered in the pollen released by the male flowers as they exit. By activating their male and female flowers at different times, Dieffenbachia encourage genetic diversity by encouraging cross-pollination using pollen from different plants.

Unsurprisingly for a plant with such fearsome powers, Dieffenbachia has accrued significance in a spiritual and medicinal sense. In Peru and other parts of the Amazon, it is ascribed occult powers: it is known as *patiquina verde*, and is considered a powerful amulet that when placed as a potted plant at the door will deflect evil spirits, just like another plant in this book, *Aloe vera*. In Brazil, it carries the evocative name *comigo-ninguém-pode*; although this translates literally as 'nobody can with me', it is variously translated as 'no one overpowers me' or 'don't fuck with me'. The Tucuna people of the upper Amazon used *D. seguine* to make curare, a poison for tipping hunting arrows.

One of the least understood and most mysterious applications of Dieffenbachia is as a contraceptive and a tool for sterilisation. The scientific literature on this plant is littered with reports of Caribbean peoples chewing the stems as a way of inducing a short-lived sterility. In the 1930s, German doctors Gerhard Madaus and

Friedrich Koch set about testing the use of Dieffenbachia extracts as a method for sterilising rats. When the Nazi SS Reichsführer Heinrich Himmler learned about this work in 1941, he ordered more research. The Third Reich wanted to find ways to carry out the mass sterilisation of concentration camp prisoners: a slow genocide that would be less time consuming than surgical castration. Records were destroyed at the end of the Second World War, but testimony at the Nuremburg Trials suggests that sterilisation experiments using Dieffenbachia extracts were carried out on prisoners, but no planned mass sterilisation programme followed. The difficulty of providing enough heated greenhouses to raise sufficient plants must have been a factor, but some reports also suggest the greenhouses containing the plants were bombed. Either way, getting hold of enough plants proved impossible.

After reading all this, you may wish to echo the emotions of my favourite houseplant blogger, Plants Are The Strangest People, who concluded a post about the darker side of Dieffenbachia by saying: 'All these years, you think you know a plant, and then you find out something like *this*. I feel like saying to the plant, *I don't even know who you **are** anymore*.' But that is the point: 'The truth is the greatest weapon we have,' as H. G. Wells put it. By educating ourselves about Dieffenbachia's history, we learn how plants can be weaponised as well as used for good.

CARE GUIDE

Light Dieffenbachia tend to grow in forest clearings so enjoy good light levels; they will take some morning sun, especially in winter. If your plant is looking miserable, lack of light is the first thing to consider, but increase light exposure gradually to avoid sunburn on the leaves.

Temperature A minimum temperature of around 15°C (59°F) is required; do not expose plants to cold draughts.

Water Steady moisture when the plant is in active growth is ideal. Dieffenbachia work well in a self-watering pot, or water when the top half of the substrate is dry (a clear pot really helps with observing moisture levels).

Humidity A minimum of 50% humidity will help your plant thrive.

Pests and diseases Spider mites love a stressed Dieffenbachia.

Substrate Dieffenbachia enjoy a fairly humus-rich substrate, but adding a chunky component such as leca or orchid bark will help to improve drainage and aeration.

Propagation Given their manner of increasing in the wild, it is not surprising that sections of Dieffenbachia stem can easily be propagated, provided they contain one or more nodes. Tip cuttings also work well. Most plants can be divided fairly easily.

Feeding Use a foliage houseplant fertiliser regularly when the plant is in active growth in spring and summer.

Other maintenance tasks As plants mature, stems will gradually lose their lower leaves and become bare canes: if you do not like this look, cutting canes back in spring will promote new growth.

Danger signs Do not be alarmed if a newly bought plant drops leaves – the same can happen to a plant that has done well all summer when winter comes. Masses of yellowing leaves are usually a sign of too much water at the roots.

Toxicity Toxic to humans and pets: most poisoning cases involve children and animals, so do not grow this if you have nibblers in your home, and take precautions with young visitors. Wear gloves when handling, repotting and pruning, and avoid touching your face; if you do get sap on your skin, wash it off immediately.

Display A forest of Dieffenbachia with contrasting foliage patterns looks stunning, and also helps the plants out by creating a more humid microclimate. You can either plant them in one trough or a large container, or keep them in separate nursery pots but combine them into one cachepot and pack the empty spaces with grit or leca.

Cultivars There are many modern cultivars to choose from: ones with leaves that resemble camouflage material are particularly popular, such as 'Reflector' and 'Cheetah'. If you are not a fan of variegation, 'Green Magic' has dark green leaves with a cream midrib. Choose a dwarf cultivar such as 'Compacta' if you prefer something that will not need regularly hacking back.

Also try . . . If you like the look of Dieffenbachia but cannot get them to thrive, try Aglaonema. These South-east Asian aroids are easier to grow than Dieffenbachia and can cope with more shade.

Dionaea muscipula

VENUS FLYTRAP

Droseraceae

Imagine the home of the Venus flytrap: your mind probably conjures up a steamy jungle, full of lush vegetation that envelops your every step. In the centre sits *Dionaea muscipula,* the air pulsing with fat flies about to be lured to their doom. Wrong, wrong, wrong. This is a plant that lives on the margins of bogs populated by evergreen shrubs and pine trees. *Dionaea muscipula*'s native range is the coastal bogs of North and South Carolina in the US, within a few dozen miles of the port city of Wilmington. Flytraps require unobstructed sunlight and plenty of moisture. They are used to a shallow water table caused by an impenetrable layer of organic matter known as a hardpan that prevents drainage, leaving the ground constantly moist. The soil is a mix of peat and sand, acidic and nutrient-poor, and flytraps are often found growing in hollows, or on sphagnum moss.

Their other requirement is fire. Flytraps do not grow under dense vegetation: if they begin to be swamped by leaf litter and overhanging plants which block their light, they will decline. Naturally occurring wildfires burn off this competing vegetation. While fires often scorch away the top growth of the flytraps, this does not kill the plant, which regrows from a rhizome – a thickened stem that stores water and nutrients, and escapes damage as it grows well below the soil's surface. Scientists have found flytraps fare best in places where wildfires occur every few years.

Flytrap habitats are biodiversity hotspots, home to other rare species such as the snakemouth orchid (*Pogonia ophioglossoides*), and fellow carnivores: the North American pitcher plants *Sarracenia flava* and *S. purpurea*, the thread-leaved sundew (*Drosera filiformis*) and the roundleaf sundew (*D. rotundifolia*).

The flytrap is teetering on the edge of extinction in the wild: it is listed as vulnerable on the International Union for Conservation of Nature's Red List of threatened species. Current estimates put the number of flytraps left in the wild at less than 500,000, just 2% of their estimated original population size. There are multiple threats: encroachment by building development, agriculture and logging, wetland drainage and continued poaching. It is now a felony to poach Venus flytraps in North Carolina, but it is a difficult crime to detect, as thousands of plants can be taken from the ground in a short time. As human development closes in, there's another issue: humans suppress wildfires to protect their property, which allows more vegetation to shade out the flytraps. Wildfires come less often, but are more severe, killing off flytrap rhizomes for good. Rising sea levels due to climate change pose a further risk, as the landscapes flytraps inhabit are often only 2-4m above sea level.

Let's smash another flytrap myth now. Flytraps don't catch many flies in the wild: the majority of creatures caught are ground dwellers. Ants and spiders each account for a third of their diet, while grasshoppers, beetles, springtails, mites, millipedes and centipedes make up most of the rest, plus the occasional flatworm when traps become submerged during flooding. Depending on which scientific study you consult, flies make up just 1-18% of the flytraps' haul. Trapped creatures dissolve into a nutrient-rich liquid the plant can absorb to make up for the poverty of its soil. There is, however, one creature that turns the tables on the flytrap: the Venus flytrap cutworm moth, whose cappuccino-coloured, faintly striped caterpillar feeds exclusively on *D. muscipula*. The flytraps' white flowers are held on stems 15-35cm above the traps, seemingly to stop their pollinators from becoming dinner too. There is scant research into their pollination habits, but sweat bees, longhorn beetles and checkered beetles seem to be the main candidates.

Dionaea is one of two species that employ snap traps formed from modified leaves. The other plant with this adaptation - the waterwheel plant (*Aldrovanda vesiculosa*) - is an aquatic plant with a wider geographic range, but its 2-4mm long traps are harder to observe. The mechanism of the *D. muscipula*'s trap itself is not fully understood, but scientists agree that it evolved out of its close relative, the sundew: the tentacles of the sundew have evolved into the trigger hairs that arm the inside surface of each clamshell-shaped modified leaf of a flytrap, usually a trio in a triangle shape. The rows of 'teeth' (or cilia) that enmesh as the trap closes seem to have evolved from sundew tentacles too.

Scientists are still delving into how flytraps attract their prey. What seems clear is that they release a cocktail of at least 20 chemicals that mimic the scent of fruit and flowers; some studies suggest that flytraps display a pattern of ultraviolet light absorption that helps prey home in on the traps. How does the trap know the difference between a potential meal and a gust of wind? Any one of the trigger hairs must be stimulated twice or more in a period of around 20 seconds for the trap to snap shut. The two sides of the leaf snap together in a third of a second, but the trap only seals tight shut if the hairs continue to be triggered by struggling prey. Once this happens, digestive glands lining the inside begin to release an acidic liquid full of proteins and enzymes to break down the meal. Each hair trigger sends an electrical signal to the whole trap, causing it to release stored energy and rapidly switch from its open concave shape to a convex shape, like an umbrella blowing inside out. Recent research has shown that these signals also produce magnetic fields: one of the first times this has been detected in the plant kingdom.

How did the indigenous people who lived alongside the flytraps understand and interact with these plants? Available information is scant: unsurprising for a culture that has been effectively erased from the landscape in the last 300 years. The Cherokee people who lived in what is now called North and South Carolina had their land taken by European settlers in the 1700s, lost their lives to introduced diseases, and were forcibly removed to so-called Indian Territory in Oklahoma.

The Cherokee name for the flytrap is *yú:gwil*, but this term was used to refer both to *D. muscipula* and *Sarracenia purpurea*, the North American pitcher plant, as well as other species. Cherokee Nation member and ethnobotanist Dr Clint Carroll told me, 'As with most Cherokee words, the name describes physical properties of the plant, so this name has been used for numerous different species that have similar appearances or "do" similar things in their environment from the perspective of Cherokee speakers.' American ethnographer James Mooney, writing in 1891, reported that Cherokee fishermen chewed a piece of what he called 'yugwilu' and spit it onto the bait and hook, to imbue them with the seductive power of the plant. This corresponds with the words of J. T. Garrett, a member of the Eastern Band of the Cherokee from North Carolina, who notes in his book *The Cherokee Herbal: Native Plant Medicine from the Four Directions* that the root of the plant was 'chewed and spit on the worm or bait'. Garrett writes that the flytrap is classed as a special medicine that was 'just not talked about to anyone'. Indigenous plant knowledge was – and still is – prized, and therefore not dispensed lightly: as Dr Carroll puts it, 'The adage "knowledge is power" has a literal meaning for Cherokees.'

The man who first brought the Venus flytrap to world attention was Scottish-born politician and landowner Arthur Dobbs, who lived in Carrickfergus, on the island of Ireland. Dobbs set out for the US in 1754 after Britain appointed him governor. In 1759 he wrote to British naturalist Peter Collinson, 'We have a kind of a Catch Fly Sensitive which closes upon anything that touches it . . . I will try to save the seed here.' The Venus flytrap first made landfall in Britain in July 1768, in a shipment of moss-wrapped plants imported by William Young, an American plant collector. This prompted feverish excitement from British naturalists hoping to get their hands on a specimen. Dobbs' friend Collinson wrote, 'I am ready to Burst with Desire for Root, Seed, or Specimen of the Wagish Tipitiwitchet.' Writers – specifically male writers – have expended a huge amount of energy exploring whether 'tipitiwitchet' – a word ascribed various meanings including a twitching fur stole, a cassava squeezer and a trap for catching rabbits – was a euphemism for the female sex organs. The flytrap's glistening red interior, fringed with tooth-like hairs, does certainly conjure up the myth of the 'vagina dentata', as Freud dubbed it – the tale told across many cultures that cautioned men of the 'perils' of sex with unknown women.

I suspect this tells us more about eighteenth-century sexual mores than it does about the flytrap. But it does point to one aspect of the plant's pull on our consciousness: *D. muscipula* 'wagishly' slipped the bounds of classification – is it male or female? Flora or fauna? Amphibian or land dweller? Intelligent or mindless? In 1773, British naturalist John Ellis gave the flytrap the scientific name *Dionaea muscipula*. Ellis claimed that this was due to 'the beautiful appearance of its milk-white flowers, and the elegance of its leaves', but the name has caused much confusion, because Dione refers to several different people in ancient mythology, including the mother of the Roman goddess of love and sex, Venus, and the mother of the Greek equivalent, Aphrodite, or Aphrodite herself. Added to that, *muscipula* does not mean flytrap in Latin, but mousetrap. And yet, somehow the name Venus flytrap has persisted.

Names also proved a problem for Carl Linnaeus, the Swedish botanist who came up with the system of naming plants and animals that scientists still use today, when he was sent a specimen of the plant. It was dead on arrival, so he could not see the plant in action, but he still deemed that the mere idea of a plant that ate animals was 'against the order of nature as willed by God', deciding that the plant must be merely allowing insects shelter from rain. Charles Darwin was a flytrap fan, calling it 'one of the most wonderful plants in the world' although he was frustrated that he could not keep one alive for long. Darwin's experiments showed that flytraps were indeed carnivorous, and resulted in his book *Insectivorous Plants*, published in 1897.

Despite its tiny size and inability to catch anything larger than a tiny lizard, artists have consistently portrayed flytraps as man-eaters. In 1880, Sir Arthur Conan Doyle published 'The American's Tale', a story about a man eaten by a giant flytrap in Arizona; the trope was reworked in the twentieth century with John Wyndham's 1951 book *The Day of the Triffids* and Audrey II in the 1960 film *Little Shop of Horrors*. And yet, there are signs that *D. muscipula* may be able to help us, not harm us. Early research into genetically modifying flytraps using a certain type of soil bacteria has shown this can boost their existing capacity to produce phytochemicals with the power to tackle human antibiotic-resistant bacteria. If we can protect these plants in their wild habitat, who knows what other secrets they may yet yield?

CARE GUIDE

Light Provide as much sunlight as you can muster. It is entirely possible to grow flytraps in an unheated greenhouse in the UK's temperate climate: they die back over winter, resprouting from their rhizomes come spring. Many Dionaea experts believe plants require this period of dormancy to thrive, during which they experience shorter daylight hours and lower temperatures, but the advent of cheap, full spectrum growlights used 12–14 hours a day have allowed some growers to experiment successfully with growing plants indoors all year round without a dormant period.

Temperature If you are growing them indoors without growlights, give flytraps a cooler resting period over winter, placing them in a brightly lit unheated room, porch or similar where temperatures drop to around 10°C (50°F).

Water Place pots in a tray of water 2cm or so deep from spring to autumn, and keep them moist but not wet in winter. The safest option is to use rainwater or distilled water. Regular bottled water and water from a home filtration system aren't suitable, but if you have a reverse osmosis (RO) system installed, or can source RO water from an aquarium shop, this is fine to use.

Humidity Venus flytraps love humidity.

Pests and diseases The two main possibilities are red spider mites and aphids.

Substrate It is still widely assumed that flytraps require a peat-based substrate to thrive, and half-peat, half-sand is often recommended. Given peat's status as a non-renewable resource, growers have been working hard to come up with a peat-free substrate for flytraps. Readymade mixes are available from UK growers Floralive and Wack's Wicked Plants, while other growers grow in sphagnum moss, or a mixture of fine milled bark, lime-free horticultural grit and perlite, to a ratio of 2:1:1. A 2cm deep mulch of sand is recommended by some growers.

Propagation Don't buy the commonly available 'grow your own flytrap from seed' kits – they usually fail as the seed needs to be sown fresh. Offsets can be carefully prised away from the parent plant and potted up separately, or cut the rhizome into pieces in spring, making sure each section has some leaves and roots, and pot up separately. Leaf cuttings using a whole leaf pulled away from the plant and laid on damp substrate will grow into new plants, but this method is slow. Flowering stems can also be propagated in this way: help to speed along the process by encasing the tray or pot in a clear plastic bag to ensure steady air humidity.

Feeding Some growers advise providing extra food for flytraps – rehydrated freeze-dried bloodworms, available from aquarium shops, for instance – but this is not essential.

Other maintenance tasks
Growers disagree over whether flowers should be removed, to nudge the plant to putting its resources into producing more traps, or left.

Danger signs Traps rapidly dying back in summer can be caused by a lack of moisture or use of tap water.

Toxicity No known toxicity to pets or humans.

Display Flytraps do best in glazed pottery or plastic containers.

Cultivars Flytrap breeders have had a field day coming up with all kinds of variations on the flytrap theme, from cultivars 'Akai Ryu' and 'Bohemian Garnet', which are a dramatic dark red all over, and the extra-large 'La Grosse à Guigui' and 'B52' to oddities such as 'Crispy Sun', which has traps with fused teeth.

Also try . . . There are many fascinating carnivorous plants suitable for indoor cultivation: sundews and butterworts act like living fly paper, while the tropical pitcher plant species in the genus Nepenthes are also widely available and relatively easy to grow indoors.

Epipremnum aureum

DEVIL'S IVY

Araceae

The island of Mo'orea is just ten miles across; on the map, it looks like a dinosaur's footprint stamped in the expanse of the South Pacific. Two million years ago, this mountainous tropical landscape was formed by the eruption of a volcano, along with the other islands in the *Tōtaiete mā* archipelago, named the Society Islands by Captain James Cook in 1769. Yet it took until the twenty-first century for Mo'orea to be identified as the one true home of a houseplant both mundane and strange. How can a plant be both mundane and strange? Excuse the pun, but the devil's ivy is in the details.

The first time I saw a mature *Epipremnum aureum* was in the Glasshouse at Wisley, the Royal Horticultural Society's flagship garden in Surrey, England. I stopped at a pillar that stretched towards the roof 12m above me; it was covered in a dense thicket of dark green oval leaves, splashed with gold and slashed with irregular cuts. Here and there, I could see stems as thick as my bicep snaking around their support. Could this monster of a woody liana be the same species as the ubiquitous houseplant with heart-shaped foliage the size of my palm? A look at the label confirmed it: *Epipremnum aureum*. Mind blown.

Some plant species make successful houseplants because they are diminutive enough to never outgrow their space on a side table or windowsill, but *E. aureum* takes a different tack. Like other aroids, its growth pattern and leaf size dramatically change as they mature, but when confined to a pot, they stay in their juvenile state. This species' genus name comes from the Greek words *epi*, meaning upon, and *premnon*, which translates as trunk: botanists call it a facultative epiphyte, meaning it can either start life in the ground or nestled in a tree branch, but will adapt its growth habit according to the landscape around it. In places where it is surrounded by trees, *E. aureum* will take advantage of the trunks as a way of hoisting itself into the canopy in search of more light; elsewhere, it simply covers the ground with abandon. Once mature, the leaves can reach 1 m or more long, and change from heart shaped to oval, with distinctive, irregular incisions to the margins. Masses of aerial roots grow from the stem, serving to anchor the plant to its host, and penetrate the soil as another way of accessing water and nutrients.

If you come across a mature *E. aureum*, examine it for flowers. If you find any, take lots of photographs and call your nearest botanic garden. Botanists describe devil's ivy as 'shy-flowering'. Given the last record of *E. aureum* blooming happened 60 years ago, you can see why. This species is the only member of the aroid family found so far that will not readily produce flowers even when mature. It took until 2016 for a team of American, Chinese and Taiwanese researchers led by Chiu-Yueh Hung and Jie Qiu to work out that this reluctance to reproduce sexually was due to the plant's lack of gibberellic acid. When they treated the cream and green cultivar 'Marble Queen' with a solution of this plant growth hormone, flower buds appeared seven to eight weeks later.

The last known reports of flowering specimens come in a cluster in the early 1960s. In 1962, American botanist Monroe Birdsey reported finding devil's ivy flowering in locations in Puerto Rico and Fairchild Tropical Garden in Florida. In 1964, Indian botanist Caetano Xavier dos Remedios Furtado published a paper on the species, noting that a plant nurtured by Mrs R. R. Sarathee on the third-floor balcony of her flat in Singapore had flowered in September 1961. The paper includes a photograph of the plant, with Mrs Sarathee standing alongside. It is interesting to note that it is planted in a kerosene tin and looks less than 1 m tall, although the leaves are larger than those of the average indoor plant – Furtado does not say, but perhaps this was a cutting of a more mature plant? The flower itself is fairly unremarkable: a corncob-like spadix (flower spike) surrounded by a spathe – a sheath-like bract. The cluster of the last known instances of flowering in a short period in the early 1960s led at least one botanist, Alan Herndon, to wonder whether devil's ivy could be subject to a phenomenon known as gregarious or mass flowering, where every plant from a particular species blooms at the same time. Could we suddenly see the world's devil's ivies burst their buds in the next few decades? I'd love to be alive long enough to find out.

Given its reluctance to flower, how has devil's ivy spread around the world so rapaciously from its origins on one tiny South Pacific island? It is able to adapt to many different settings, which is one of the reasons why it makes such a good houseplant, but this has been less welcome in some regions. It has become an invasive species in many tropical climates, smothering whole stands of indigenous plants and trees with its lush growth. It is likely that the answer lies in *E. aureum*'s ability to reproduce very readily in a vegetative fashion, plus its adaptability to a wide range of conditions. The smallest section of discarded stem will quickly grow into a new plant; it has presumably relied on humans carrying it across oceans, and people have planted it as reliable greenery for tough conditions in subtropical and tropical climates.

Despite its ubiquity, we have not understood that much about devil's ivy until recently. The minutiae of flowers – the number and structure of stamens and so on – has traditionally allowed taxonomists to delineate one species from another, although genetic research is helping refine and redefine the boundaries in recent years. I do not have space to lay out the tortuous and torturous route devil's ivy took to reach its current scientific name, but it boils down to this. It was first given the name *Pothos aureus* in 1880, by French and Belgian botanists Edouard André and Jean Linden, who claimed that the specimen they described had come to Gant from the Solomon Islands. This legacy remains today, as one of its most popular names, golden pothos, despite the fact that it looks very different from the actual species in the genus Pothos, another type of aroid. A move to another aroid genus, Scindapsus, occurred in 1908, until the flowering incidents in the 1960s finally allowed for a more accurate reassessment of its place in the aroid family; after a brief spell in Rhaphidophora, devil's ivy finally ended up in the genus Epipremnum.

Since its first designation as a Pothos back in 1880, the species has been recorded as coming from the Solomon Islands, east of Papua New Guinea, but no wild specimens were ever found there. Aroid expert and taxonomist Peter Boyce finally solved the mystery when he compared *E. moorense* from Mo'orea to the familiar devil's ivy: they were identical, but for the former being plain green. Boyce suspects that the golden variegated form we know was a horticultural selection of the nineteenth century: a time when plant hunters were not always honest about the provenance of their finds.

Epipremnum aureum began to be distributed as a houseplant from the 1920s, reaching peak popularity in the 1950s and 60s: its star faded in the following decades, at least in terms of its reputation in the horticultural press: in his 2001 book *Potted*, British garden designer Andy Sturgeon noted, 'People really love this ugly plant, perhaps because it's easy to grow. They seem to do really well in chip shops.' As we have already seen with the Aspidistra, people are generally immune to such snobbishness, and devil's ivy has continued to spread its vines around homes, offices and indeed chip shops across the world. Like several other species, it earned many common names, including money plant, hunter's robe and, my favourite, centipede tongavine. It's still often called golden pothos, and also mistaken for another Epipremnum species, *Epipremnum pinnatum*, as well as another similar-looking aroid, *Philodendron hederaceum*.

Perhaps one stumbling block for this species has been its shy-flowering nature. This is not because the flowers are particularly exciting, but because its lack of flowering has proved problematic to breeders, who are usually able to cross-pollinate different plants to create new cultivars and hybrids. Since the turn of the twenty-first century, there has been a relative explosion of new devil's ivy cultivars, partly as a result of the expanded breeding possibilities brought by tissue culture, and partly due to the use of radiation treatments to induce genetic mutations. One other boon for devil's ivy came in the form of research from the 1990s and 2000s that showed *E. aureum* was successful at removing pollution from indoor air, although as we have discovered from the spider plant chapter (see page 37), the importance of these findings has often been overblown. However, more recent work by scientists at the University of Washington in Seattle, published in 2018, found that genetically modifying *E. aureum* with a gene from a rabbit boosts its ability to break down the toxins benzene and chloroform. The species' ability to thrive in a wide range of light conditions made it the perfect plant to test out the idea that plants could improve air quality. And this species may have a few more tricks up its sleeve – more research is under way to employ devil's ivy's ability to treat wastewater, provide an eco-friendly way to tackle infestations of the highly destructive Indian white termite and slow down damage to bridges during floods.

CARE GUIDE

Light Unless you shove devil's ivy in full sun next to your cacti and succulents, it is hard to give it too much light in the average home, provided your plant is given time to adapt gradually. Although these plants can cope with your least auspicious spot, in darker corners the leaves may revert to plain green and the stems will become spindly.

Temperature 15°C (59°F) and above will be fine, and plants can cope with short spells down to 10°C (50°F).

Water The leathery leaves allow the plant to cope with periods of drought, but it will do better if you water it once the top half of the potting mix is dry.

Humidity Humidity of 40% or more will be fine.

Pests and diseases If kept in good condition, devil's ivy is not prone to pests, but mealybugs are the most likely to appear.

Substrate I have seen plants in rude health growing in soil from someone's garden; that said, best practice is to pot into a substrate that allows for good drainage. Adding a quarter or a third of bark chips and leca or perlite to regular houseplant potting mix, John Innes No. 2 or similar should give your plant its best life. Devil's ivy also does well in hydroponic and semi-hydroponic setups.

Propagation Stem cuttings will strike easily in a glass of water, in damp sphagnum moss or leca. Make sure you cut a piece that contains at least one node – the point where a leaf joins the stem. And for obvious reasons, if you find *E. aureum* seeds for sale online, they are fake.

Feeding Some people never feed their devil's ivy to no obvious detriment, but for the best growth, use a foliage houseplant feed regularly when the plant is in active growth.

Other maintenance tasks Plants that go out of bounds or start looking spindly should be cut back; avoid doing this in the depths of winter, but any other time is fine. Give plants a regular rinse down or wipe with soft water to remove dust.

Danger signs If you leave your plant sitting in water, the leaves will rapidly turn yellow and drop. Straw-like patches on the leaves indicate that you have moved your plant too quickly into full sun.

Toxicity Toxic to pets and humans.

Display If you want your devil's ivy to produce larger, more mature leaves, give it something to cling and climb onto, rather than letting it trail: this could be an obelisk, a trellis or a hoop, or you can let it climb a wall.

Cultivars The best-known cultivars are the cream-splashed 'Marble Queen', and the lime green 'Neon', although a dozen or more have been introduced in the last few years, many of which are hard to tell apart. Smaller-leaved cultivars 'Pearls and Jade' and 'Njoy' are irregularly marked in white and green, while 'Shangri La' has twisted foliage and gold splashes.

Also try... There is a whole array of vining aroids to try, including *Rhaphidophora tetrasperma*, which is often sold as the mini Monstera, and the slightly trickier to grow *Scindapsus pictus* or satin pothos, whose juvenile leaves are patterned with silver.

Ficus lyrata

FIDDLE-LEAF FIG

Moraceae

If I had to choose one defining image of the late 2010s houseplant aesthetic, it may be a *Ficus lyrata* in a rattan basket, its uppermost leaves brushing the ceiling of a whitewashed room. The fiddle-leaf fig seems such a poster plant of the twenty-first century indoor garden that it's hard to imagine its past. But fiddle fever broke out long before people were snapping them on smartphones. In her 1939 book *The Indoor Gardener*, American garden writer Daisy T. Abbott was extolling the virtues of a 'new' type of rubber plant (*sic*) 'with large, shining leaves shaped like a violin' that looks 'most handsome against the usual plain background of the modern home.' As Dorothy Jenkins and Helen Van Pelt Wilson wrote nine years later in their book *Enjoy Your House Plants*, 'Here indeed is the perfect tall plant, with clean, straight lines for the light, empty corner of the library or sun room.'

These two houseplant books were written half a century after the first specimens of *F. lyrata* were collected from their home in West Africa in 1890 by German botanist Otto Warburg. In 1903, Sander & Sons, a nursery based in St Albans, England, displayed the fiddle-leaf fig at the Ghent Quinquennial Exhibition in Belgium, earning a first-class certificate. In 1911, the *Gardeners' Chronicle* profiled the species under the heading 'new and noteworthy plants' reporting that 'this plant has since become popular, particularly in the United States, on account of its decorative qualities and its good nature under cultivation'.

Let's just clear up confusion about the name; or, at least, present a clear account of the mess. In the English language, this species is known as the fiddle-leaf fig, referencing the leaves, which, though variable in shape, tend to start narrow at the point nearest the stem, widening out to make a pear shape that could, with some artistic licence, resemble a violin. The accepted scientific name, *F. lyrata*, refers to the leaf as being shaped like another, somewhat more obscure musical instrument, the lyre. In my opinion, the leaf is pear-shaped. But I have also come across references to the leaves being banjo-shaped, and to the name *lyrata* referring to the designs marked onto German lye bread. This sounds particularly unlikely, given that the lye in this case refers to the alkaline solution the bread is dipped in before baking. To deepen the confusion further, you may find the fiddle-leaf fig wrongly sold under the scientific name *F. pandurata*, which really does mean fiddle-shaped: the trouble is, this name belongs to another fig species that comes from China, not Africa. Phew.

The plant's native range extends across the coastal nations of tropical West and Central Africa, from Guinea-Bissau to Gabon, via Nigeria, Liberia, Ghana and six other countries. It is an evergreen tree that can reach around 12m tall but is often smaller, with a trunk clothed in pale grey bark lightly scored with vertical striations. It grows in lowland rainforests, where it is warm and wet: temperatures range from 20°C (68°F) to 30°C (86°F). It is also cultivated as a shade tree both in Africa and in regions with similar climates, including Florida in the US, Cuba and the Philippines.

Ficus lyrata is sometimes described as a strangler fig. The 750–900 (depends on which Ficus scientist you ask) species in the genus include a wide range of growth habits; from soaring trees to epiphytes, vines and shrubs. Strangler figs begin life as epiphytes that sprout in the nooks and crannies of trees where birds, bats or other consumers of figs deposit seed-laden faeces. They gradually extend roots towards and into the ground, and can eventually kill off their host tree. Several popular houseplants, including the curtain fig (*F. microcarpa*), weeping fig (*F. benjamina*) and the rubber plant (*F. elastica*) count among their number. Indeed, the Khasi and Jaintia people in the Indian state of Meghalaya utilise the aerial roots of the latter to make living root bridges to cross rivers. Is *F. lyrata* a strangler fig, then? It depends on who you ask. One authority on Ficus reports that the fiddle-leaf fig starts life rooted in the ground, whereas another claims it sometimes begins life as an epiphyte, before becoming freestanding as it matures and roots into the soil. I have seen no evidence for the latter position, but given the large number of Ficus species with a similar growth habit, it cannot be completely disregarded.

Ficus lyrata does share one fascinating habit with the rest of the genus: its method of reproduction. If you're wondering why you have never seen fig flowers, that is because they are hidden from view. An immature fig is, put crudely, a bunch of inside-out flowers: botanist Scott Zona helpfully described it to me as a green flask or urn lined with tiny, reduced flowers – just the male, pollen-producing stamens and the female pistils. These are often called enclosed inflorescences or syconium, but for simplicity we can just call them fruit. *Ficus lyrata*'s fruits resemble small green apples dotted with yellow marks; like the rest of the figs, they are produced in abundance and are eaten by a wide range of creatures, including monkeys, bats and parrots. An indoor potted fiddle-leaf fig is unlikely to produce fruit, but even if it did, the seeds inside won't be viable unless you happen to live in the tree's native range. Via a chemical signature we do not yet understand, each Ficus species attracts a few – often just one – species of extremely short-lived wasps from the family Agaonidae. The tiny wasp that is the sole pollination partner of *F. lyrata* is *Agaon spatulatum*: without this creature moving its pollen from flower to flower, the fiddle-leaf fig's seed does not become fertile. No other flying insect can do the job.

The female wasps are just 1–2mm long, but they must squeeze their way inside the fig via a special scale-lined hidden entrance called the ostiole: this sits at the other end of the fruit from its stalk. The manoeuvre is quite a feat: wasps often lose wings and antennae in the crush. Once inside, the female lays her eggs inside some of the female flowers, spreading pollen she has brought in from another fig as she goes. Her work over, the female dies. Her wasp larvae develop, eating the seed as they grow, and flowers without eggs develop into seeds. The wingless male wasps hatch first: they are fated to live an extremely short life spent entirely inside a fig. They cut a hole in the seed coat where the female wasp larvae remain and fertilise them before they have had a chance to emerge, then chew a hole in the side of the fig, so that once the fertilised females are ready, they can exit and fly away with a payload of pollen to find a new fruit to pollinate. Once the females are gone, the fruit will finally ripen to become attractive to its eaters, who spread the seed that has not been parasitised by the wasp to create new plants.

There is much that scientists still do not understand about coevolved mutualism, their name for the complex, interdependent relationship between each fig tree species and its wasp pollinator species, but there are concerns that climate change may impact populations of either the tree or the wasp, which could be devastating for the many creatures that rely on figs for food. In places such as Hawaii where the fiddle-leaf fig grows as an introduced tree, it is not currently a threat because the trees' pollination partner does not dwell there, and therefore cannot spread viable seed. Were *Agaon spatulatum* to spread to those areas, this tree has the potential to become a problematic species that could crowd out native species.

For the moment, though, *F. lyrata*'s attempts at world domination centre on the houseplant world. As Steve Kurutz wrote in the *New York Times* in 2016, 'This decade belongs, undeniably, to the fiddle-leaf fig. Open the latest issue of *Elle Decor* or *Architectural Digest* and you will likely spot a fiddle-leaf fig.' Houseplant 'influencers' flooded social media with images of their figs, such as Hilton Carter, whose tree Frank became an Instagram star in its own right. These plants were more than living decor; they were on a par with a pet. Perhaps Michelle Slatalla, gardening editor at interior style website Gardenista, was right when she said fiddle-leaf figs' large, glossy leaves are similar to the big round eyes of a baby.

Social media is awash with advice on growing fiddle-leaf figs: some of the advice, such as cleaning the leaves with beer, is just plain wrong, but viral videos of people shaking their trees may not be a complete red herring. Research into other trees such as Liquidambar has concluded that gently shaking the main stem for as little as 40 seconds a day may result in a thicker trunk and a more compact tree. But the overwhelming problem faced by fiddle-leaf figs grown indoors is lack of light. Recreating the light of the fig's African home is not that easy: they need to be positioned close to the largest window, or in a conservatory, not shoved in the far corner of the room in the only space available. Not enough light slows the plant's processes including transpiration, where the plant pulls up water from bottom to top, drawing water from the roots and passing it into the cells to fuel photosynthesis and keep cells turgid. A lack of transpiration means the potting mix dries out slowly, and roots are at risk of rot. Growers think they are overwatering, when really the root (excuse the pun) of the problem is not enough light.

You may have never seen one, but a purple-leaved *F. lyrata* exists. In 2013, Jietang Zao and a team of scientists from the University of Florida published a paper describing creating a genetically engineered fiddle-leaf fig with dark purple leaves. Thanks to the addition of grapevine genes to its DNA sequence, the foliage is packed with anthocyanins, the pigments that turn green leaves red or purple. Don't necessarily expect to see a purple fiddle-leaf fig in a garden centre near you soon: another paper published in 2019 reported issues with defoliation and reduced disease resistance in the transgenic figs, and regulatory hurdles mean that this plant is unlikely to make it to the open market any time soon. Still, it will be fascinating to see whether and how genetic modification of houseplant species begins to take the industry in new directions.

CARE GUIDE

Light This tree needs light, and lots of it. A room with floor to ceiling or roof windows, or a warm conservatory, is ideal.

Temperature Although *F. lyrata* can cope with occasional drops in temperature down to 5°C (41°F) or so, your plant will be far happier at a range of 18–30°C (64–86°F). Avoid cold draughts and don't place it close to heating outlets.

Water See the note above about the relationship between water and light: when the fig is in active growth, it should be watered once the soil is dry on the surface and barely damp around the rootball itself.

Humidity Dry air will stress out an already stressed fiddle-leaf fig: aim for 50%.

Pests and diseases Red spider mites just love a fiddle-leaf fig, especially one that is living in substandard conditions, but it can also suffer from any of the other major houseplant pests, particularly mealybugs.

Substrate Regular houseplant potting mix will work, but if you tend to be generous with the watering can, add extra drainage material, such as leca, perlite or grit.

Propagation A single leaf placed in soil or water will very rarely grow a stem, but stem cuttings can be taken. If you want to make your fiddle-leaf fig bushier at the same time, remove the top section, cutting just above a node (the point where the leaf stalks join the stem) and trimming the cutting to just below a node. Air layering (rooting a cutting while still attached to the plant) is another approach.

Feeding There are specialist fiddle-leaf fig fertilisers available, but a general feed for foliage houseplants works just as well, applied when the plant is actively growing.

Other maintenance tasks Turn your fig regularly to prevent it becoming lopsided as it will lean towards any light source.

Danger signs Dull, drooping leaves are a sign that the plant is either in need of water, or the roots are rotting due to waterlogging. Check the rootball immediately. Red or brown dots on leaves – particularly new ones – are a sign of oedema caused by irregular watering.

Toxicity Take care when trimming, propagating or pruning or if the tree is damaged, as the milky latex produced can irritate skin. Toxic to pets.

Display Given its size, *F. lyrata* is usually a floor-standing specimen. Some growers place large pots on wheels so they can move them around to enjoy the best possible light. Consider underplanting the bare soil around the base with a low-growing or trailing plant such as an asparagus fern (*Asparagus densiflorus*), a tough leathery-leaved Peperomia such as *P. angulata*, or even a bunch of air plants (Tillandsia) rested on the surface.

Cultivars A more compact cultivar of the species, 'Bambino', is now available, which has smaller leaves and maxes out at around 1 m. Variegated specimens have also inevitably come on the market, sating the current appetite for cream-splashed foliage: these are trickier than the plain-leaved species, so probably best left to those who have already mastered fig care.

Also try . . . For those who have tried and failed with the fiddle-leaf fig, the ponytail palm (*Beaucarnea recurvata*) and the Dracaenas *D. marginata* and *D. fragrans* will tolerate less light and more irregular watering.

Hedera helix

ENGLISH IVY

Araliaceae

Where time has obliterated all other traces of plantlife, pollen grains remain. These microscopic reproductive cells possess a super-hard outer shell made of sporopollenin, a biological polymer. It is pollen grains that confirmed English ivy's long association with humans, well before the concept of houseplants existed: before houses existed, for that matter. Even though ivy is a relatively poor pollen producer, pollen records from Neolithic settlements around Britain show enough evidence of its presence to suggest that people were gathering its boughs during the Late Stone Age, possibly even earlier, as fodder for animals – most likely deer. When fresh grass was scarce in winter, evergreen ivy was an alternative food source. It's a practice whose traces are still found in pockets of rural Britain and Europe, where farmers feed sick sheep ivy as a pick-me-up.

Since then, humans have been using English ivy in all sorts of ways. In the medieval period, ivy aphid, the greyish bugs that feast on the plant, were crushed to create a skin-coloured dye used for painting illuminated manuscripts; in the 1800s, leather cutters sharpened their knives on the roots; fishermen used resin from the cut woody stem as a bait lure. Ivy had medicinal applications too. In Nicholas Culpeper's *The Complete Herbal*, published in 1653, it was considered a treatment for everything from the plague to a stitch. Even today, ivy is an ingredient in expectorant cough mixtures. Within living memory, *Vickery's Folk Flora* reports that an ivy leaf placed between the toes was reputed to rid the feet of corns, while a stew of ivy leaves cleaned the blue serge suits of railway workers. This last application isn't so far-fetched. Ivy leaves contain saponins, surfactant chemicals that foam when placed in water and that have traditionally been ingredients in soap.

Another of the suggested uses for ivy in Culpeper's *Herbal* was as a cure for drunkenness and hangovers. English ivy's history has been entwined with that of the grapevine (*Vitis vinifera*) as far back as records go. The ancient Roman god of wine, Bacchus – Dionysus to the ancient Greeks – was depicted wearing a crown of ivy leaves, and carrying a branch of ivy, or in later depictions, the thyrsus, a stalk of giant fennel (*Ferula communis*) topped with ivy leaves. Wearing an ivy wreath on the head was thought to stave off the ill effects of wine, as could drinking from a goblet made of ivy wood. British pub names such as The Ivy Bush and The Ivy Leaf offer an echo of the tradition, brought to the island by the Romans, of hanging ivy outside drinking establishments. Roman understanding of ivy wasn't completely tied up in mythology. Roman author Pliny the Elder makes mention of *H. helix* in his work *Naturalis Historia* (Natural History), an encyclopaedia of the natural world, noting several different varieties and their growth habits, including one with variegated leaves.

Ivy leaves are also littered through early Greek and Roman texts – not as a matter for discussion, but in the form of a piece of early punctuation known as the hedera: literally, a small ivy leaf understood to indicate a break in the copy (*hedera* simply means ivy in Latin). This was handy, because Roman text tended to be crammed in with little regard for how it might be read. In the absence of other typographical marks we now use to make sense of a text, the hedera served as a useful indicator for a pause, or sometimes as a filler for blank space. Typographical experts haven't pinpointed an exact reason why an ivy leaf was chosen for this purpose, but it certainly suggests that ivy leaves were one of the most recognisable leaves of all – something that remains true today. Even though the hedera gradually lost traction in typography over the centuries, many font designers still include the hedera in their work.

Hedera helix is sometimes confused with poison ivy (*Toxicodendron radicans*), the rash-inducing American native. *Hedera helix* is a woody liana of the temperate forests of Europe, its range stretching from Britain and Ireland to Ukraine, and extending as far south as Turkey. Estimates of how tall it can grow vary, but 30m is a reasonable estimate: that's roughly the height of a seven-storey building. Young, spindly stems the width of a strand of wool can over time become woody vines 25cm across. It's able to thrive in a wide range of soil types, but does best in moist ground.

English ivy has earned itself a bad rap throughout history for a variety of alleged crimes: damaging buildings onto which it clings, becoming an invasive species in various US states and killing the trees it uses for support. Let's look at the items on *H. helix*'s charge sheet in turn: it can damage buildings, but it tends to inveigle its way into existing cracks rather than compromising sound brickwork. A cloak of ivy can in fact bring benefits to buildings, protecting them from frosts and damp like a warm, waterproof duvet. And invasiveness? Like so many plants, ivy was taken to North America with colonists, and has choked indigenous woodland plants in some states. But it is *H. hibernica*, often given the common name Irish ivy, that is responsible for most of these issues. To confuse matters, *H. hibernica* is considered by some botanists as a subspecies of *H. helix*. Either way, it's safe to say that the *H. helix* cultivars grown as houseplants tend to occupy the tamer end of the invasiveness scale. And finally, although ivy often clings to tree trunks, the vine draws no nutrients or water from them: it is not a parasite. A large mass of ivy in an already weakened tree can amplify the effect of winds and hasten its fall in stormy weather, but it also offers a shelter for wildlife and food for invertebrates and birds.

How does the ivy scale obstacles with ease? First, roots start to grow over the surface of their support, then they grow a covering of tiny root hairs that exude a glue-like substance. The root hairs lodge themselves in tiny crevices, then dry out into a spiral shape that helps to pull the surface into even closer contact with the vine. Ivy also has a strategy to cope with both the darkness of the forest floor and sunlight at the top of the tree canopy. Just like another plant in this book, *Monstera deliciosa* (see page 123), English ivy fundamentally changes its appearance with age: botanists know this as heteroblasty. For the first ten or so years of life, the juvenile form has leaves with three to five lobes that are adapted to life in the shade, at the base of a tree and climbing its trunk. The seedlings are skototropic, growing towards the dark, not the light, because dark means a tree trunk whereas light means an open space with nothing to climb. The adult form kicks in as the ivy reaches the top of whatever it's been clinging to – tree, house or wall – and thus emerges into the light. It is only then that flowers will appear; the leaves are different, too, losing their lobes to become heart shaped. The ivy we grow as houseplants is almost invariably of the juvenile form: it's naturally suited to conditions indoors, being smaller and able to cope with lower light levels. As this transformation is triggered by light, age of the plant and complex hormonal changes, our housebound ivies tend to remain in their more dainty juvenile state, however long we have them.

When did ivy first begin to be brought indoors as a houseplant? By medieval times boughs of ivy featured in European Christmas celebrations, alongside its fellow European evergreen, the holly (*Ilex aquifolium*), but its classical associations with gods made it a sometimes ambiguous presence in Christian symbolism. It wasn't until the Victorian age that ivy reached the zenith of its popularity as a houseplant. It was far cheaper and faster growing than the other key plants of the era, the Kentia palm and Aspidistra: and in Europe and North America, you could simply take a cutting from the nearest wall or tree. Ivy served a further purpose: it could be grown to clothe a room or to mask the next door neighbour's brick wall or other less than thrilling vistas. As Tovah Martin wrote in *Once Upon a Windowsill*, 'Ivy was a plebeian plant. It grew by leaps and bounds. Anyone could obtain it easily and cheaply, and anyone could maintain it without deep immersion in the study of the horticultural arts.' *Hedera helix*'s needs really matched the cold, damp, draughty homes most Victorians had – plus it could cope with the air pollution from coal fires and gas lighting that wiped out so many other houseplants.

Victorian garden writer (James) Shirley Hibberd's beautifully illustrated monograph on ivy documents various ways the plant was grown indoors, suggesting that for those who had failed to grow ferns in glass cases, ivy was an achievable goal. Hibberd notes that French and German homes were particularly rampant with ivy, but it was prominent in British homes too. Ivy grew around window frames, picture frames and door frames; it romped up indoor trellises that arched over settees; it spilled over pots and embraced statues. The Americans were not to be left out either: the *American Garden* noted in 1882, 'Kept in a shady corner of the room, we may extend its friendly wreaths around our picture frames, or train them where we will and they resent it not.' *Hedera helix* plunged in popularity as a houseplant once central heating became the norm – it does not cope with dry air and the thermostat set to a steady 20°C (68°F): red spider mites and crispy leaves often follow.

CARE GUIDE

Light Direct summer sun will burn indoor ivy. Other than that, ivy will adapt to most spots in the home, but if your plant starts to struggle and throw out undersized leaves, that is a sign it needs more light. Variegated leaves can also turn plain green when light levels are low.

Temperature Ivy hates central heating, especially at night. Keep it well away from radiators, fires and other heat sources during winter; if possible, move to a light, cool room such as a porch, conservatory or unheated bedroom.

Water A steady supply of moisture is preferred, but don't allow the substrate to become waterlogged. Slow down watering in winter but don't allow the substrate to become bone dry.

Humidity At least 50% humidity will keep your plant in a happy state.

Pests and diseases Red spider mites are the main threat. If your plant looks miserable, this is the first thing to check. Aphids often colonise young, tender growth tips.

Substrate A soil-based potting mix such as John Innes No. 2 is ideal.

Propagation Ivy in its juvenile form will easily root as stem cuttings.

Feeding Use a feed for foliage houseplants diluted by at least half every time you water when the plant is in active growth.

Other maintenance tasks When grown indoors, *H. helix* inevitably becomes straggly, so be diligent in regularly pruning back any weak stems: these can be propagated and added back into the pot to make a bushier plant.

Danger signs Pale, stippled leaf undersides usually indicate red spider mites or thrips. Crispy leaf tips or whole stems are usually a result of dry air.

Toxicity Toxic to humans, cats and dogs.

Display One of my favourite illustrations from one of my favourite houseplant books, Ladybird book *Indoor Gardening*, published 1969, shows a plain green English ivy growing out of an old teapot. *Hedera helix* is one of the few plants I can countenance planting in a novelty container: channel the Victorians and let it wreath a picture frame or ornament, or simply trail it from a head pot in the manner of a wild head of hair.

Cultivars There are dozens of smaller-leaved cultivars that are ideal for growing indoors: silvery variegated 'Glacier' and the bird's-foot-shaped 'Needlepoint' are both classics.

Also try . . . If a houseplant has the slightest resemblance to *H. helix*, you can bet someone has added 'ivy' to its common name. Swedish ivy (*Plectranthus australis*), kangaroo ivy (*Cissus antarctica*) and wax ivy (*Senecio macroglossus*) are easy 'ivies' to try.

Howea forsteriana

KENTIA PALM

Arecaceae

When Queen Victoria died on 22 January 1901, her staff enacted a twelve-page set of funeral instructions she had compiled ready for the occasion. Many of her requests broke with traditions of the time, and some certainly seem peculiar to modern readers. The British monarch had been wearing black mourning dress for four decades since the death of her husband Prince Albert, but she insisted on being buried in a white dress, with the white lace veil she last wore at her wedding over her face as a shroud, and she stipulated that mourners at her funeral should not wear black. Tucked around her body were many surprising items, including a plaster cast of her late husband's hand and his dressing gown. But the Kentia palms which stood at either end of the coffin would have been a surprise to no one.

These large palms dominate the photographs of Queen Victoria's coffin lying in state in the Albert Memorial Chapel at Windsor Castle. Their fronds arch over the bearskins of the four Grenadier Guards who stood on duty, heads bowed, but they are not noted in many newspaper reports of the scene. I suspect this was because palms in general were such a fixture of upper class life at the turn of the twentieth century that they did not warrant a mention. As professor of design history Penny Sparke noted, palms acted as a frame for the cluttered interior of Victorian homes. 'They intro-duced an exoticism of the tropics, as well as memories of empire and of an untamed world in which nature held sway over culture,' she wrote. Queen Victoria grew Kentias at her royal residences, society weddings were held underneath living arches made of two Kentia palms, and Kentias filled the lobbies of luxury hotels.

For the Victorians, the main boon of the Kentia was that it

could thrive in domestic settings as well as heated glasshouses. Like the Aspidistra, it could withstand the dark and often draughty conditions of people's homes and did not keel over when exposed to fumes from gas lighting as so many other plants did. Fully grown plants were a huge luxury, though. Kentia palms could eventually reach 3m or 4m tall when potted indoors, but they are slow growing, and seed germination was erratic and sluggish, which meant these plants were time consuming and thus expensive to raise, taking up space in a heated greenhouse. The *New York Sun* in 1893 reported that a large Kentia palm could cost $50 – the equivalent of around $1,500 today. One advantage was that they stood up well to being handled, and to being treated as cut foliage, which was a boon because these palms did not always stay in one spot for long. If you could not afford a whole plant, florists would be happy to sell you a cut frond or two to decorate your dinner table. If you had a little more money, you could rent out a whole plant for special occasions. Those who did have the means to own one would send them to 'board' at a florist's greenhouse for a month or two every year, presumably to recover from the stress of living in a dark Victorian home.

More than a dozen different palm species were part of the indoor plant trade in the 1800s, but the Kentia palm rose to pre-eminence fairly quickly after its introduction in 1871. Or, more accurately, I should say Kentia palms. There were two different species in the Howea genus, and the Victorians grew both indoors: *Howea forsteriana*, also known as the thatch palm, and *H. belmoreana*, the curly palm. (In the palm trade these names were shortened to 'Bells' and 'Forsters'.) They come from Lord Howe Island, a crescent-shaped speck of ancient, eroded volcano jutting out of the expanse of the Tasman Sea between Australia and New Zealand roughly six miles long and one and a quarter miles across at its widest point.

Why, then, is this a profile of *Howea forsteriana* and not *H. belmoreana*? The two were roughly neck-and-neck in the popularity stakes in the Victorian era, albeit used in slightly different ways. The slightly taller *forsteriana* was often placed as a floor specimen in the corner of rooms, whereas the shorter *belmoreana* looked good on a table. Somewhere along the way, *forsteriana* won when it came to indoor cultivation, although *H. belmoreana* is still available from specialist palm growers and grown as an outdoor palm in warmer climates. The reason is that *H. forsteriana* tends to be that bit easier – and faster – to grow.

What are the differences between the two species when growing on Lord Howe Island? *Howea forsteriana* has drooping leaflets and tends to grow on alkaline sandy soil, known as calcarenite, at lower altitudes on the island, often near the beach. *Howea belmoreana* is slightly shorter, and its leaflets tend to point up, and this palm only grows on dark brown volcanic soil of neutral or acid pH, often at higher altitudes. The other key difference is that although both are wind pollinated, the two species flower six to seven weeks apart. Botanists puzzled over the fact that such a small island could be home to two similar yet distinct species of palm, which appeared to have a common ancestor – a rare phenomenon known as sympatric speciation. How had two species living cheek by jowl evolved in this way? The answer was in the soil. DNA research comparing the palms' genes showed that they share their ancestry with an Australian palm called *Laccospadix australasicus*; the two Howea species separated 1.92 million years ago. The nature of the soil it occupied and the accompanying fungi in that root zone seem to have shifted the flowering time of *H. forsteriana*, which also experienced more stress as a result of increased exposure to salt and onshore winds. But that is what makes it such a great houseplant: the climate of Lord Howe Island is classed as warm to cool subtropical, with temperatures averaging 20–25°C (68–77°F) in summer and 14–20°C (57–68°F) in winter, which is more akin to average room temperatures. Their tolerance to shading from taller trees, drought stress and salt spray means they can readily adapt to the confines of life indoors in a pot.

Lord Howe Island was never home to an indigenous population of humans, but around 47% of its indigenous plants and 60% of its indigenous insects are unique to the island. It was first occupied by a small number of whalers and women from South Pacific islands who had married them in the 1830s. They survived by selling supplies to the whaling boats that anchored there periodically. The first dwellings were thatched with the fronds of *Howea forsteriana*, hence the common name thatch palm. As the whaling industry began to wane, another source of income arose to fill the gap. In 1869, Charles Moore, director of the Sydney Botanic Garden, visited the island to examine its flora. Moore collected seeds from both palms and sent them to Baron Ferdinand von Mueller, curator of Melbourne's Royal Botanic Gardens. Mueller named them *Kentia forsteriana*, after William Forster, a New South Wales politician, and *K. belmoreana*, after the Earl of Belmore, who was governor of New South Wales at the time. Florentine botanist Odoardo Beccari moved them to the genus Howea in 1877. By February 1872, London nursery Veitch's was offering *H. forsteriana* plants for sale, and soon other nurseries in Belgium and the Netherlands were selling both plants and seeds to enthusiastic buyers. The islanders began to collect seed to sell by the bushel to nurseries all around the world, using jute or canvas straps to shimmy up the palm's slender trunks. They removed the 'fingers' – long spikes of seeds – and took them down the tree for shelling and packing into bags ready for export.

The death of Queen Victoria did not see the end of the Kentia craze. *Howea forsteriana* was the species that graced every hotel's palm court, where a light orchestra played classical music for the upper classes to enjoy as they indulged in afternoon tea. There were Kentias on the *Titanic* when it sank in April 1912; Kentias in the palm court at the Ritz hotel in London when it opened in May 1906, and potted Kentias for sale in the famous London department store Harrods. This was a worldwide phenomenon rather than simply a British penchant: a report on Kentia palms in the *Los Angeles Times* in 1929 noted: 'They are used in hotel corridors, on banquet tables and for gala occasions – in Tokio [*sic*], Manila, Singapore, Suez, Cairo, Paris, Barcelona, Berlin, London, New York and Los Angeles.'

Kentias – first the seed, and later the plants themselves, which were raised in a nursery set up in the 1970s – are still the island's major export, although since the end of the Second World War tourism has become another significant income stream. But something nearly put paid to this success story in 1918, when the supply steamship SS *Makambo* was grounded just off Neds Beach in the north-east of the island. It was refloated nine days later, but in the meantime the vessel's population of rats abandoned ship and rapidly occupied the island. The population of *Rattus rattus* (surely the best scientific name ever) reached huge proportions in just a few years and developed a taste for Kentia seed; annual collection rates dropped from 4,500 bushels in 1919 to 877 bushels in 1925. A bounty of 6 pence a rat resulted in a dent in the population as islanders shot as many as they could, and various species of non-indigenous owls were introduced as rat predators, although they in turn caused a drop in the island's own owl species. The Kentia seed harvest recovered, but the battle against rats has continued well into the twenty-first century. In 2019 a $16m rat-eradication programme by the Lord Howe Island Board saw 42 tonnes of poison-laced bait pellets and 22,000 lockable traps distributed around the island. The plan proved controversial with islanders who were worried about the poison's impact on other wildlife, but it was declared a success: the last rat was reportedly despatched that same year, although two more rats were spotted and killed in 2021.

Although Kentias are now grown in nurseries around the world, Lord Howe Island still makes an income from its iconic palm, selling young seedlings to buyers globally. And although the Kentia palm's heyday was undoubtedly in the late Victorian and Edwardian eras, it has been a steady feature of the houseplant scene ever since; a statement plant with a hefty price tag, but one that should last for generations in the right spot.

CARE GUIDE

Light Kentia palms can survive dark corners but will hardly grow at all and will eventually start to suffer. Avoid direct sun, but choose a brightly lit spot.

Temperature Down to 12°C (54°F) will be tolerated, but ideally temperatures should be 16°C (61°F) to 28°C (82°F).

Water When the plant is actively growing, you can water liberally provided the substrate is right (see below) but make sure you do not leave water sitting at the base of the pot.

Humidity Kentia palms cope better than most palms with dry air but will still protest with brown tips to their leaves. Keep away from heat sources such as radiators and fires.

Pests and diseases Spider mites, scale, mealybugs, thrips.

Substrate Palms resent root disturbance, so only repot when they are truly rootbound. A deep pot is required, but only use a pot slightly bigger than the old one. Use bespoke palm potting mix or John Innes No. 2, adding some perlite, grit or leca if you are the type who tends to be too generous with the watering can.

Propagation *Howea forsteriana* can be grown from seed, but you may be waiting several years to achieve a decent-sized specimen. Plants are often sold as a number of young seedlings in one pot, so that a fuller looking specimen can be achieved in less time. These plants can be separated out by teasing apart the different root systems and potting them up individually.

Feeding Apply a foliage houseplant fertiliser every couple of weeks when the plant is in active growth.

Other maintenance tasks Wiping fronds down with a damp cloth is time consuming; if plants are manoeuvrable enough, take them outside in summer or to the shower in winter to wash them down.

Danger signs If seedlings are crowded into a pot, one or more of them may die off completely; check whether the plant needs repotting, or if the roots are waterlogged. Blackened leaves can also be a sign of overwatering. Leaves turning straw like or brown can be a sign of sunburn: you can put your plants outside in summer, but start them in deepest shade and gradually increase the amount of light they are getting.

Toxicity No known toxicity to pets or humans.

Display This statement plant deserves a spot where you can enjoy its form without the distraction of other greenery: but bare soil at the base of mature specimens could be clothed with a trailing plant such as the frothy *Pilea libanensis*, or the drought-resistant rabbit's foot fern, *Davallia tyermanii*.

Cultivars There are no widely available cultivars of either species of Howea, although red-stemmed and variegated forms have been found on Lord Howe island.

Also try . . . If you only have room for a smaller palm, the parlour palm *Chamaedorea elegans* tends to stay under 2m when grown potted, even when mature. The lady palm, *Rhapis excelsa*, is another slow-growing palm with some interesting – if expensive – variegated cultivars.

Hoya carnosa

WAX PLANT

Apocynaceae

I'd like to start a special club for readers of this book. How will we know each other? By the Hoya blooms we wear.

Weddings are the sole twenty-first century occasion when one may be expected to sport a buttonhole, but for Victorian men and women, particularly those of the upper classes, they were more of a regular affair. Hoyas sometimes featured: *Exchange & Mart's* 'Housekeeper's Room' column of 1876 suggested several Hoya-based flower combinations for buttonholes, including blue nemophila and lily of the valley; cornflowers; and a red carnation with a spray of maidenhair fern.

I had better not lay it on too thick here: Hoyas were not the most popular choice of flowers for buttonholes. But that may be down to a botanical peculiarity of the genus: namely, they are produced on persistent peduncles – a flower stalk that doesn't drop off after blooming. On *Hoya carnosa*, these peduncles look rather like the burned end of a tiny cigarette; they are held on a short piece of stem that emerges from the point where the leaf joins the main stem, and the longer they are, the older they are. If you cut off a peduncle, the plant has to grow a new one. This slows down re-flowering, so cutting them off was a sacrifice that may have deterred growers from displaying their Hoya blooms as buttonholes too often, and pushed up the price at the florist. But I just love the idea of showing off my Hoyas to the world in this way.

Today, *H. carnosa* is enjoyed as much for its foliage as its flowers: particularly prized are leaves with lots of 'splash', an unscientific term for a phenomenon common across many Hoya species, a form of variegation that marks leaves with a random pattern as if they have been splashed with silver paint. This is known as blister variegation, as explained in the *Begonia maculata* chapter (see page 21).

There is an ever fluctuating number of species in the genus Hoya, partly because some have not been known to science until recently, and partly because taxonomists – scientists who place plants into categories – are still wrestling over their names: in some cases whether they even belong in the genus. So, forgive me if I remain deliberately vague in saying that the number falls around 350. The species' geographical range extends across Asia including India, Japan, the Philippines, Thailand and Vietnam, as well as Australia. *Hoya carnosa* is indigenous to southern China, Japan, Laos and Taiwan, and is probably the best known of the genus, as well as being the 'type' species: it was one of the first Hoyas to be identified by western plant hunters during the mid- to late 1700s as they fanned out across the planet, 'discovering' things that were probably already well known to the people who lived there.

Specimens of *H. carnosa* first arrived at Kew Gardens in 1802, but it was not until 1810 that it was described to science by Scottish botanist Robert Brown, best known for explaining Brownian Motion, the movement of microscopic particles in a liquid or gas. He named the plant genus Hoya in honour of his friend Thomas Hoy, head gardener at Syon House, the west London home of the Duke of Northumberland; there are no known records of whether Hoy was pleased with such an honour. In a very few years, the plant was firmly established as a popular 'stove plant' – a vine of heated greenhouses – and gradually worked its way into homes across the world in subsequent decades.

The species dwells in the edges of woodland in southern Japan, Taiwan and central to southern China, and flowers from late spring to early autumn. It grows as an epiphyte, colonising nooks and crannies in trees and twining around other vegetation to secure a foothold and a greater share of the light. *Hoya carnosa* seeds are designed to float on the wind, with fluffy appendages that allow them to fly up far into the canopy and lodge in any suitable spot.

Epiphytes often make good houseplants, as they do not require a huge root run to succeed: even a large *H. carnosa* can thrive in a pot that will fit in most domestic settings. What it does like is something to cling to, be it an obelisk, wire hoop or bamboo cane. Like other Hoya species, the thick, leathery leaves are designed to hang onto other vegetation, or even another stem of the same plant, something you may have experienced if you have ever tried to disentangle a Hoya from a trellis. By growing at a 90° angle to the leaf stalk, which itself is at 90° to the main stem, the foliage can hook onto other leaves with ease. If no support is immediately to hand, the whole stem can move around in search of an anchor. Brown noted in his 1874 *Manual of Botany* how the tip of an *H. carnosa* stem will move in a circle of around 1m in search of support. New Hoya owners are often confused by the way that their plants produce long whippy stems without leaves: this is the plant's way of prospecting for new ground. The leaves tend to follow on once the stem has detected enough light and a place to anchor itself; this is why it is important to offer Hoyas support.

Once the flowers start to form, a cluster of thin stalks called pedicels emerge from the peduncle, each topped with a five-sided shell-pink bud. Many a Hoya owner has kept vigil over these tightly clasped buds, waiting for them to burst open, firework-style. The flowers have earned the species many common names, including wax plant (although this is also thought to be inspired by the waxy leaves) and porcelain flower. The Victorians were hot on the symbolism of plants and flowers, and according to the 1825 work *Floral Emblems* by Henry Phillips, Hoyas represented sculpture, because the leaves look like works of art rather than nature. The outer corolla is a shell-pink, fuzzy five-pointed star (think starfish); inside is another star shape, this time a darker pink and waxy, made up of the fused male and female sex organs of the plant.

The flowers open at night, programmed to emit their perfume to appeal to their pollinator partners, which are nocturnal moths: their pale colouration makes the flower clusters shine out in the darkness. Until recently, scientists didn't know how Hoyas in general, and this species in particular, were pollinated. From 2013 to 2015, Ko Mochizuki and colleagues staked out *H. carnosa* plants growing on Amami Ōshima Island in Japan at night to record which creatures visited the flowers. The wax plant's flowers produce copious amounts of clear nectar, with production peaking at midnight, earning the species another of its common names, the honey plant. As the moths traverse the flowers in search of the nectar sitting around the 'skirt' of the waxy central star, they slip onto twinned pairs of pollen sacs, which are conveniently located ready for such a pickup. Each pair of pollen sacs is held by a cliplike tissue structure called a corpusculum, which attaches to the moth's hairy feet or legs. When the moth moves on to the next flower, the pollen sacs detach to allow successful pollination to occur, leaving the corpusculum behind on the moth. By counting the numbers of corpuscula with missing pollen sacs, the researchers discovered that the wax plant's main pollinator is the owl moth (*Erebus ephesperis*), one of the largest moths with a wingspan of 9-15 cm. As the common name suggests, humans tend to see the face of an owl in the patterning of cream and brown that is designed to confuse predators by breaking up its outline.

Humans also found *H. carnosa* nectar useful. In 1816, John Maher informed the London Horticultural Society that the nectar-filled blooms were a useful addition to hothouses raising grapes, as insects that troubled the grapes would be distracted by the Hoya nectar, thus leaving the precious fruit alone. And in 1851 the *Cottage Gardener* called it '"the honey plant" of our boyish days' and claimed this clear fluid was 'said to be the wine of the heathen gods'. Even today there is a well-populated Facebook group called the Hoya Lickers Society. Yes, I have licked my *H. carnosa*'s flowers, but I couldn't possibly recommend that you try it: indeed, you may be better off smearing it on your face. French cosmetic research company Silab has isolated a yeast from the floral nectar and cultured it to create a skin treatment that, it claims, improves the skin barrier by rebalancing microbiota of the skin.

Modern-day witches ascribe magical powers to *H. carnosa*, which they call pentagram flower: the blooms are dried and worn as an amulet, according to Cunningham's *Encyclopaedia of Magical Herbs*. Extracts from the leaves have traditionally been used to treat chronic infections of the middle ear in Bali, and scientists have begun to trial this as an alternative to the antibiotics usually used in treatment of the condition. Other traditional uses for *H. carnosa* leaves include topical treatments for skin rashes, acne and boils.

After huge popularity in the Victorian age, Hoyas spent most of the twentieth century as solidly reliable plants for the heated greenhouse or home. In 1931 the *Rural New Yorker* was calling this plant old-fashioned: its ease of growth and ability to live for decades perhaps discouraged nurseries from offering plants, figuring that most households, if they wanted a Hoya, already had one.

Contemporary houseplant growers have rediscovered this plant in a big way, and I couldn't be happier. So, next time your wax plant flowers, pop a peduncle or two in your buttonhole and I'll look out for you.

CARE GUIDE

Light These plants can take some direct sun, especially in winter, but protect them from being hit by direct summer sun. Leaves that have been exposed to sun may take on a reddish tinge, known as 'sun stress' – too much sun will bleach the leaves or cause blistering.

Temperature This is one of the most cold-tolerant Hoya species, able to cope with temperatures down to around 12°C (54°F).

Water This species copes with periods of drought, helped by its succulent, leathery leaves. But it will grow better if watered regularly when in active growth. Wait until the top half of the pot is completely dry and the bottom half just damp before watering thoroughly. In winter, keep the soil dry and only water when the leaves start to look deflated.

Humidity The wax plant's leathery leaves mean dry air is not too much of an issue, but it may well grow more rapidly when humidity is 50% or more.

Pests and diseases Mealybugs are the main pest, and infestations can be hard to eradicate completely. Aphids can also strike, targeting young tender growth in particular and causing the plant to produce distorted leaves.

Substrate Bear in mind the plant's tendency to grow as an epiphyte – it is used to a chunky substrate that drains freely. I tend to use a soil-based mix (John Innes No. 2 or 3) and add a handful of fine orchid bark or leca. Many people grow Hoyas successfully in pure leca or pon.

Propagation Stem cuttings are the norm, but experiments with cuttings made of single leaves with the leaf stem (petiole) attached have sometimes been found to produce whole plants in time. If you want to try growing Hoya from seed, make sure the flowers get pollinated: either by rubbing successive flowers with a fine paintbrush or cotton bud. Once seedpods form, wrap them in a plastic or paper bag to catch the seeds as the pod pops.

Feeding Hoyas will benefit from a regular high potash feed in spring and summer when the plant is in flower: either use it at half strength, or alternatively use orchid feed.

Other maintenance tasks
If variegated cultivars start to produce large amounts of pure white leaves, remove the stem as this can weaken the plant.

Danger signs If leaves start to soften and turn yellow, or drop one after another, check for root rot caused by waterlogging. Plants may abort flowering when they are too dry.

Toxicity None known.

Display Mature Hoyas require plenty of support: mature plants need a substantial obelisk, trellis or staking system. Clear plastic pots allow you to assess the plant's watering needs accurately, and can be hidden inside a cachepot of your choice.

Cultivars There are hundreds of different cultivars: one of the most famous include 'Krinkle 8' with bumpy leaves, 'Krimson Princess' with creamy centres and green margins, and 'Krimson Queen', which has green centres and creamy-white margins; the latter two exhibit a pink tinge to the foliage. Silvery-leaved cultivars such as 'Wilbur Graves' and 'Grey Ghost' are among the most prized (read: expensive), although they are particularly slow growing. 'Compacta', a cultivar with leaves that look as if they've been squashed together, is a curiosity, but its nooks and crannies are the perfect hiding place for mealybugs.

Also try . . . There are dozens of Hoya species now on the market, some very different in appearance and habit from the wax plant. *Hoya bella* is a good choice if space is limited, while *H. linearis* trails without need for support.

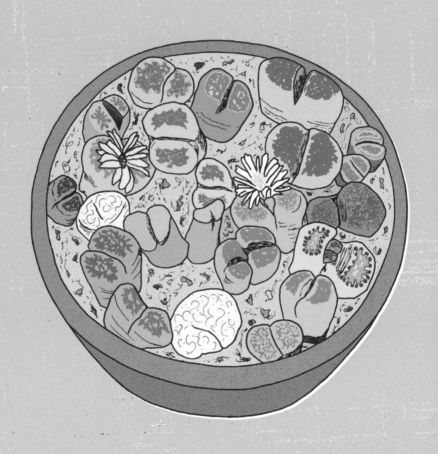

Lithops

LIVING STONES

Aizoaceae

What do you see when you look at a Lithops? A pebble, a set of cow's hooves, a face or maybe even a pair of buttocks? Stone plants are the Rorschach test of the plant world: enigmatic little succulent lumps that fascinate botanists and houseplant growers alike.

The scientific name for the genus comes from the Greek *lithos* meaning stone, and *ops*, meaning face, but collectors often call them belly plants, because in order to see Lithops in their native habitat, you must be flat on the ground. Rather than singling out one of the 38 or so species, this chapter examines living stones as a genus: largely because diverse species are often grown together as houseplants, and no single one dominates the indoor plant industry.

Part of the appeal of Lithops is that, when you grow them in a pot, you can place them at eye level to enjoy every subtlety of their patterning and texture, which is often tricky in the wild. Each species is precisely coloured and intricately patterned to match the particular geology around them, giving them superb camouflage. During the drier parts of the year, living stones conceal themselves further, shrinking down into the soil and shrouding themselves in wind-blown sand, grit and organic matter. The occasional seed capsule above ground is often the only sign of their existence.

Almost all Lithops hail from South Africa and Namibia, and are concentrated on the drier western side of the tip of Africa, with a few colonies just across the border in Botswana. This is tough territory for plants: average annual rainfall is 100–500mm (the UK gets roughly 1000–1200mm), and temperatures of 40°C (104°F) and above are fairly common. A few species dwell on dry grasslands and experience higher rainfall, but all Lithops must be equipped to endure extreme temperatures, soaring light levels and dry air. So they have evolved some equally extreme coping methods.

Their structure is simple: just two succulent leaves, fused together and joined to a carrot-like taproot that can quest far into the soil in search of moisture, as well as anchoring the plant. By staying small and round, Lithops keep their surface area to volume ratio as tiny as possible, so less water will be lost via transpiration through the plant's stomata, or breathing pores. Like all succulents, they also use a specialised form of photosynthesis called Crassulacean Acid Metabolism, where their stomata open at night rather than in the day (more on this in the *Crassula ovata* chapter, page 45). The leaves are also able to store large amounts of water. During dry spells, they will shrivel like an empty balloon, but as soon as rain falls, they swell to their maximum size, their tops rising above the soil surface like buns in an oven as the tissues take on all that extra water.

This is not without its problems, because one of the greatest risks to a colony is being eaten: humans, baboons, cattle, sheep, goats, mice, cutworms, armoured crickets and grasshoppers all munch the juicy leaves as a last resort in dry periods. To limit the damage from grazing, the plant hides its most vital parts below ground. Slice a plant down the centre and you will see a heart-shaped structure: the dip at the top is where the two leaves join, and deep within is the meristem, the growth point where new leaves and flowers emerge. Below that is a long taproot. These underground parts allow the plant to regrow if chewed.

By mimicking the terrain around them, they can stop being spotted in the first place by creatures that locate food by sight, including humans and baboons. Take *L. amicorum*, which was only named for science in 2006 and dwells in Namibia. It grows in fields of quartzite, a blue-grey metamorphic rock, matching its blue-grey tones exactly. The top of the plant – the 'face,' as it's known – is mottled with grey and red spots, dots and lines (known by Lithops experts as rubrications) that echo the iron-rich loam where they grow. South African species *L. dorotheae* is known for its blood-red rubrications against a beige background, perfect for hiding in plain sight in a land littered with both rusty red feldspathic quartzite and grey and white quartzite. The rubrications are like a fingerprint, unique to each plant, and made up of tannin-filled cells whose purpose we're really only beginning to understand. The bitter-tasting tannins probably help to make the plants less tasty, but they are also thought to serve as a sunshield.

Lithops' iceberg-like habit means not much of it is exposed to the sun. This can be a good thing, preventing radiation damage to plants. But living stones also need to photosynthesise when completely obscured by debris. Scientists are still figuring out how Lithops carry out this balancing act, but here is what we know. Most plants carry the equipment of photosynthesis – the chloroplast cells – just under their epidermis or skin: in Lithops, they are held deep within, made accessible by translucent areas called leaf windows that channel light inside. In some species these windows cover most of the top surface of each leaf; in others, mini-windows are scattered about. This phenomenon is common to various succulent plants, including another species profiled in this book, *Curio rowleyanus*, where they look like a dark green stripe across each of the pea-shaped leaves (see page 51). Leaf windows can be hard to spot: hold a pot of Lithops up in front of a light, and look for areas of flesh resembling frosted glass.

Lithops also deploy a suite of chemicals that act as a sunscreen, filtering out harmful radiation while letting through the light needed for photosynthesis. This consists of the tannins mentioned above, as well as flavonoids. And calcium oxalate crystals – found in many houseplants – are thought to help to scatter light and direct it down into the nether regions of the plant's tissue. Lithops have one other trick up their sleeves to deal with too much light: non-photochemical quenching. Molecules in chloroplast cells that find themselves burdened with more light than they need for photosynthesis start to vibrate faster, so excess energy is dissipated as heat.

There is one time of year when it's easy to spot Lithops: for a few days every autumn and winter, usually after a spell of rainfall, they burst forth a yellow or white flower from the centre of the leaf pair in a bid to attract as many insects as possible. The daisy-like blooms open in the afternoon and shut at night, repeating this daily until they are pollinated. Some species have lightly scented flowers: some say this smell resembles mimosa, others call it honey-like. The seed pod that follows pollination is ingeniously designed to only open when conditions are right. Once dried, it resembles a Trivial Pursuit game counter, with five to eight seed compartments arranged like slices of pie. Each chamber has an independently operated lid that stays shut to protect its precious cargo. The lid opens only when it is hit by a water droplet, exposing the dust-like seed to the rain. The seed will be flooded out of the chamber with the hope of finding a suitable spot to germinate. Any seeds that do not make it out of the pod are locked under the lid once more as the capsule dries out, ready for the next rainfall. Once flowering is over, the plant puts out a new pair of leaves: these emerge from the central fissure, set at right angles to the original pair. The existing leaves begin to shrivel as the plant reclaims its resources, allowing space for the new leaves to grow. The leaf husks stay in position for some time, acting as a protective cloak to the new growth.

It was not until the twentieth century that Lithops joined the ranks of popular houseplant species, although the first records of a westerner stumbling across living stones happened in 1811. English botanist William John Burchell recorded his trip to the region in his book *Travels in the Interior of Southern Africa*. The journey did not start well. His convoy of wagons got separated as they travelled across Northern Cape province at night, and they were scratched by thorn bushes as they tried to find their way. Having finally re-united with the rest of the party, Burchell ignored warnings from the indigenous Khoekhoen people who were his guides, and attempted to take specimens of the thorny plant – an Acacia – that had caused him so much trouble. He became entangled once more and feared his clothes would be torn off in the struggle to escape. Two of his men cut him free. We can only speculate as to what the Khoekhoen guides made of this incident because, as is so often the case, we only get to read the colonial side of the story.

Burchell then picked up a 'curiously shaped pebble' only to discover it was a plant. He named it *Mesembryanthemum turbiniforme* and noted its mimicry of the surrounding ground. It's impossible to know from the information he recorded which exact species he saw: Lithops continued to be classed as Mesembryanthemums until 1922 when they earned their own genus name, encompassing just nine species. Around this time, Lithops began to be cultivated as plants for greenhouses and windowsills, subsequently appearing in succulent nursery catalogues. By 1932, the catalogue of Herman Tobusch nursery in Villa Park, Illinois, was offering seeds of a dozen different Lithops species and varieties. The number of recorded species has expanded, with new additions in the last few decades as more Lithops have given up their secrets to researchers prepared to lie in the dust to find them.

However, plant poaching, overgrazing by farm animals and habitat destruction have put wild Lithops populations at huge risk of extinction. *Lithops lesliei*, one of the most widespread species, has in recent decades started to be sold as an ethnomedicinal plant in the markets of South Africa, although the impact of this trade is as yet unknown. The rather descriptive Zulu name for this plant, *dumbasibumbu*, and the Setswana name, *marago-abanyana*, both mean swollen labia. South African botanists Gideon F. Smith and Neil R. Crouch note in a report on *L. lesliei* that this plant is usually marketed both as a love charm and a treatment for bleeding during pregnancy.

CARE GUIDE

Light Reserve a space on your sunniest windowsill for these plants. But take care to expose them to increased light levels gradually: even Lithops can be burned when grown behind glass and exposed to full summer sun. Good ventilation helps reduce the risk of burning.

Temperature Lithops will survive spells at 50°C (122°F) in summer, but will be happy at 20-30°C (68-86°F) in homes. During winter, move pots to an unheated room kept at a minimum of 4°C (39°F).

Water Avoid watering at all during winter while the new leaves are emerging; just mist if the stones become very wrinkled. Start watering in May, once the leaves from the previous year have completely dried out. Water when the substrate becomes dry, upping the frequency as the temperature rises: an occasional soaking is preferable to daily dribbles, but never let the pot sit in water. Halt watering in October.

Humidity Dry air is this plant's usual experience; in winter, moist air may induce rot. Do not put this plant in a terrarium or a pot with no drainage.

Pests and diseases Thrips, red spider mites and mealybugs are the most common issues.

Substrate Lithops sold by non-specialist growers are almost always potted in a substrate that holds too much water, so repot immediately. Growers tend to use a mix of loam-based potting mix such as John Innes No. 2 and drainage materials, at a ratio of 1:1 or 2:1. Some growers report great success in growing living stones in a soil-free substrate such as pumice or pon: these hold no nutrients so you will have to feed more regularly.

Propagation Raising Lithops from seed is relatively easy; mix the dust-like seed with a small amount of fine horticultural sand to make sowing easier. Sow in autumn or spring into your usual substrate, sieved to remove larger lumps, and place pots in a heated propagator held at around 20°C (68°F). Large clumps can be divided, but this takes time as Lithops is a slow-growing plant.

Feeding Feed occasionally and sparingly using specialist cactus and succulent feed during the growing season.

Other maintenance tasks
Try pollinating plants when they flower by moving a cotton bud between the flowers of two different plants to transfer pollen, repeating for a few days in a row until the flowers begin to fade.

Danger signs Pale spots can be a sign of a spider mite infestation. Wrinkled leaves can be a concern, but rot can just as easily be the cause as dryness, so check the substrate before adding more water.

Toxicity None known. Indigenous people have been known to eat Lithops as a thirst quencher; Lithops expert Steven Hammer, who has sampled them, says they taste like green pepper.

Display A mixed group of Lithops showcases the interplay of the colours and patterns of different species. However, this tends to be more difficult to maintain in the long run than individually potted species. A porous terracotta container helps to keep root rot at bay, but if you use a suitable substrate (see above), plastic or glazed china will not cause any problems. Top dress pots with a mulch that contrasts with or matches the Lithops, from neon pink aquarium grit to crushed rock. Lithops roots can be up to 7 or 8cm long, so your pot should be deeper than that: 10–15cm, ideally. Some growers successfully raise Lithops in shallow pans, but the plants may be smaller as a result.

Cultivars The most popular and easiest species to start out a collection include *L. lesliei*, rusty red *L. aucampiae* and olive-green and grey *L. olivacea*. There are dozens of cultivars of these species to explore.

Also try . . . The closely related succulents in the genera Conophytum and Fenestraria offer some equally strange plant silhouettes, such as *Conophytum burgeri*, which resembles a glossy green marble.

Monstera deliciosa

SWISS CHEESE PLANT

Araceae

I nearly crashed my car while thinking about *Monstera deliciosa* the other day. The insurance paperwork would have been deeply unconvincing: 'I was wondering whether we would ever truly know why Monstera leaves have holes, and forgot to indicate at a round-about.' Of all the iconic species in this book, this one has occupied the most space in my brain – and indeed my home – in recent years. I am not alone. The Swiss cheese plant is an emblem for all things exotic, earning its place among the charismatic megaflora, a select group of huge species including the titan arum (*Amorphophallus titanum*) and the giant redwood (*Sequoiadendron giganteum*) that catch the public's imagination in a way few other plants have: the difference is that you can grow Monstera easily at home. And yet many of the things you think you know about this species are wrong.

Let's start by dispelling the three major Monstera mixups. First: this is not a South American plant. *Monstera deliciosa* comes from Central America, specifically southern Mexico and Guatemala, although it has become naturalised in many parts of South America and other tropical countries. Second: it is not a Philodendron, even though it is sometimes given the name split leaf philodendron. It was placed in the genus at one point in history, but taxonomists have established that Philodendron and Monstera are two totally separate genera within the aroid family. Third: the leaves do not develop holes as they grow. The shape of each leaf is genetically predetermined before it unfurls, so if a leaf starts out whole, it stays whole.

The name Monstera comes from the Latin word *monstrum*, which has three meanings – a bad portent, a monster, or something that provokes fear and wonder. While *M. deliciosa* can be tamed indoors, it does live up to its name in the wild. Botanists have found plants growing in mountainous spots in southern Mexico surrounded by dry scrub and Agaves; defying the monocultures of the palm oil plantations in Costa Rica by vining up oil palms (*Elaeis guineensis*); clambering through lowland jungles 90m above sea level; and occupying cloud forests at elevations of up to 2000m. The Swiss cheese plant can grow with its roots in the soil at ground level, as a lithophyte anchored to rocks, or as a hemiepiphyte, starting off at ground level then climbing trees, or vice versa. If a Monstera can access enough light, space and water, it will become huge: leaves grow more than 1m long, stems can reach a diameter of 9cm and plants have been recorded scaling up to 20m into trees.

When Monstera hunting in the wild, the first thing you notice may not be perforated leaves, but a curtain of feeder roots. These roots are the thickness of a bungee rope, emerging from the stems to seek out the soil below to draw up water and nutrients. Shorter anchor roots keep stems clasped tight to their support. How do Monsteras find a tree to climb? Their seedlings are skototropic, meaning they are genetically programmed to be drawn to darkness, which is an indicator of the presence of a nearby tree to climb. As a seedling develops into a mature plant, the leaves change radically. The first leaves start out life small, whole and heart-shaped, but each successive leaf displays more and larger splits and perforations than the last: 101 perforations on one side of the midrib is the current record.

Windowed leaves are not a common feature of foliage, aside from the genus Monstera and a few other members of the aroid family. The process that forms them is known as programmed cell-death mechanism: long before a leaf emerges, a pattern of cells is genetically predestined to die back to create a unique pattern of holes and splits. Botanists have devised different theories over the years as to the purpose of the holes: mimicking pest damage so actual pests steer clear; withstanding high winds; camouflaging foliage from predation by herbivores; letting light or rain pass through more mature foliage at the top of the plant, to reach young leaves and roots below; and regulating temperatures in a hot climate. But the currently favoured hypothesis comes from American botanist Christopher D. Muir. Put yourself back in that forest: light is in short supply, as most of the sun's rays are blocked by the trees above. Occasionally, shafts of sunlight – Muir calls them sunflecks – penetrate the canopy. To maximise the chances of sunflecks falling onto its foliage so it can harness that solar energy for photosynthesis, Monstera has leaves that go large. By adding holes, the actual plant tissue may have the same overall area, but the leaves cover more space, increasing their chance of getting hit with a sunfleck.

This may not be the last theory to surface: research into aroids is undergoing a renaissance after decades in the doldrums. Monstera taxonomy has historically been as tangled as a Monstera's feeder roots: many species are challenging to study in the wild owing to their size and location, and even more difficult to record using the traditional method of the dried and pressed herbarium samples, because they are too fleshy and huge to process. The last major revision of Monstera was published in 1977 by American botanist Michael Madison, the result of 18 months studying the genus in Central and South America.

Now botanists are examining Monstera once again, and have managed to provide new information on a long-standing debate. There seemed to be two forms of Swiss cheese plant found in people's homes: one is larger in every way, with short, wide spaces between the nodes, and leaves that become heavily perforated with age; the other has smaller, less perforated leaves and longer, skinnier internodal spaces. The latter is often called 'small form', or *borsigiana*. Botanists Marco Cedeño Fonseca from Costa Rica and Mick Mittermeier from the US set out to study Monstera in 2019, criss-crossing Central America in search of specimens. They concluded that the smaller form of *M. deliciosa* was a separate species known as *M. tacanaensis*, first identified by Japanese botanist Eizi Matsuda from a specimen collected on the slopes of Volcán Tacaná, the second highest peak in Central America, in the early 1970s. Madison's 1977 Monstera review demoted *M. tacanaensis* to a form of *M. deliciosa*, but Cedeño Fonseca and Mittermeier found enough differences to revive the name in its own right. Whereas *M. deliciosa*'s natural range is limited to southern Mexico and Guatemala, *M. tacanaensis* is native to Mexico, Guatemala, Costa Rica and Panama, and found at elevations of 300–2500m. Their fruit differs too, *M. tacanaensis* being less fleshy and tasty than *M. deliciosa*. So, take a look at your Swiss cheese plant closely, it may not be a *M. deliciosa* after all.

While *M. deliciosa* rarely produces an edible harvest as a house-plant, its fruit is a feature of mature plants grown outdoors and in large glasshouses; they will flower at about three years old. The flowers are tiny, packed together on a spike-shaped spadix, which is hooded with a modified leaf called a spathe. During flowering, the spadix heats up a few degrees above the ambient temperature and releases a sweet scent to draw its pollinators, various species of scarab beetles and also flies. These are trapped by the enclosing spathe for about 48 hours before being released, gaining a coating of pollen as they exit. During their confinement the beetles copulate, and snack on pollen and gum exudates produced by the plant. It takes five days to complete the flowering and pollination process, and a year for the fruit to ripen. Another unknown is what species disperses the seeds of the Swiss cheese plant: could it be birds, monkeys or some other creature? This has not yet been documented.

One of *M. deliciosa*'s common names is the fruit salad plant. This is about as accurate a description as you will read, because no one seems to agree on how the fruit tastes: it is compared to pineapples, melons, bananas, coconuts, strawberries, custard apples, soursop, jackfruit and even American sweets called Jolly Ranchers. The edible part is the colour of banana flesh, but this fruit peels itself: the corncob shape is covered in small green hexagonal tiles or scales, which drop off once the fruit is ripe. It is important not to rush this process, as unripe fruit is unpalatable: the calcium oxalate crystals that are common in members of the aroid family will make your lips tingle at the very least. Even once the fruit is fully ripe, tiny black particles, which are the remnants of the flowers, can make the mouth burn.

The fruit has no doubt been eaten by the indigenous people of Central America since before recorded history, but it has not become a worldwide hit in the same way as another tropical fruit native to Central America, the avocado (*Persea americana*), perhaps because of the time it takes to ripen and the risk of an unpleasant reaction to unripe fruit. Monstera fruit is sometimes still offered for sale in tropical climates where it has naturalised, for instance in Florida, and in parts of Costa Rica where it is mixed into milkshakes and made into desserts. There are not many recorded uses of *M. deliciosa* in traditional medicine, aside from a snakebite cure involving a decoction of the root in Martinique and a treatment for arthritis made into a tea of the leaves and roots in Mexico.

Specimens of the Swiss cheese plant were first collected in 1832 in Mexico by Bavarian botanist Wilhelm Friedrich von Karwinsky, although it was Danish botanist Frederik Michael Liebmann who sent the first live plants to European botanic gardens. In 1858, plants at Kew Gardens' tropical glasshouse had borne fruit, prompting the *Gardeners' Chronicle* to print an engraving of its 'singular' form. It gradually became ensconced in the warm fug of the heated glasshouses of Europe. Every country in the world seems to have coined an imaginative common name: in Spanish *M. deliciosa* is *costilla de Adán*, and in Portuguese *costela-de-adão*, both meaning Adam's rib. In its home country of Mexico, it is *piñanona*. Another Spanish name is *mano de tigre* (tiger's paw), and Sicilians call it *zampa di leone* (lion's paw). The Brits know it as the Swiss cheese plant, while the French say *plante gruyère*. The Spanish word *ceriman* represents the whole plant but it is also widely used as a name for the fruit.

Monstera deliciosa's history as a houseplant really begins in the 1950s, when homes began to offer an amenable environment for a tropical aroid, mainly as a result of the arrival of central heating. It suited the mid-century design aesthetic perfectly: its sculptural leaves can be seen in the Eames House in Los Angeles, designed by Charles and Ray Eames. In 1952, the *Architectural Review* devoted its May issue to indoor plants with an illustration of a Swiss cheese plant on its cover; inside, the Monstera is described as 'the aspidistra of tomorrow'. The houseplant boom of the past decade has seen Monstera skyrocket in popularity again, with the leaf becoming a symbol for all things indoor garden related. Millennials are inextricably linked to the Monstera trend: leaves have featured on every conceivable item from matchboxes to gin glasses. I'm currently lobbying Unicode, the registration body for emojis, to add a Monstera emoji to the suite of pictograms: surely the perfect way to cement this species' place as a twenty-first century houseplant icon.

CARE GUIDE

Light Plants tend to get leggy and sparse in dark corners. They do best close to an eastern or western window, or a metre or so from a south-facing window, but be prepared to move your plant in summer when light is more intense.

Temperature Plants will be happiest around 15°C (59°F) and above but will cope with 12°C (54°F) in winter provided the substrate is not soggy.

Water It is better to give your plant a soak in a bucket of water every three weeks than a light dribble weekly. When plants get large, make sure you check soil at root level with a moisture meter or your finger before watering, and only water when the substrate feels only slightly moist or dry in this region.

Humidity It can cope with 40% humidity but will enjoy more if offered.

Pests and diseases This species is able to resist most pests, but stressed specimens are most likely to succumb to thrips or red spider mites.

Substrate The Swiss cheese plant would not have become as popular as it is today if its taste in potting mixes was anything other than wide-ranging. Your main task is remembering to repot: Swiss cheese plant roots fill a pot at a prodigious rate, and growth will start to stall when plants are potbound. You can put one in over-the-counter houseplant potting mix, but my personal recipe is one part fine orchid bark and leca to three parts good quality peat-free potting mix or John Innes No. 2.

Propagation Monstera is easily propagated from a stem cutting provided it includes at least one node. Larger stems will need a clean hacksaw: I leave the cut to dry out for a couple of hours, then seal it over with some softened beeswax. Often cuttings will already have grown some aerial roots: placing in a glass of water will produce a bigger root system before planting, or you can just put cuttings straight into some potting mix as described above. Stems can be air layered, also.

Feeding Monsteras are hungry feeders, so regularly apply foliage houseplant fertiliser at the suggested strength when the plant is in active growth, which may well be all year round.

Other maintenance tasks People are often terrified to cut back their plants, but this is the way to keep Monsteras within bounds. Turn plants regularly (until they get too large) to ensure even growth.

Danger signs Black patches on the leaves can be caused by plants being left in cold conditions. Watch out for straw-like marks indicating sunburn. Lots of yellowing leaves usually mean your plant is waterlogged. The creamy areas on variegated leaves almost inevitably develop brown patches unless you get the care regime exactly right. Plants looking dull with pale stippled marks are often undergoing a pest infestation.

Toxicity Leaves are toxic to humans and pets. Only consume the fruits when they are fully ripe.

Display Support is essential, but how you supply that is up to you:

moss pole, trellis, a large piece of wood or a stake can all work. The Monstera at the Eames House is displayed on a wheeled platform, supported by a large piece of twisted wood and surrounded by other tropical plants.

Cultivars The vogue for variegated aroids resulted in a price hike for cream-splashed Monsteras: a 'Thai Constellation' with yellowy-cream variegation and another with white and green variegation that is usually labelled *M. albo borsigiana* dominate the market, although other cultivars are slowly becoming available.

Also try . . . If you prefer a smaller-leaved vining aroid, *Rhaphidophora tetrasperma* – sometimes sold as the mini Monstera – is equally easy and attractive.

Nephrolepis exaltata 'Bostoniensis'

BOSTON FERN

Polypodiaceae

I have no desire to inflame any existing beef between two American cities, but it may be more accurate if this plant was called the Philadelphia fern. In 1894, Fred C. Becker, a florist in Cambridge, Massachusetts, received an order of 200 sword ferns from Philadelphia grower Robert Craig. Becker noticed something different about one of them: rather than the stiff, erect bearing of the rest, this specimen grew faster but had a more relaxed form, with wider, more arching fronds.

Becker realised this was something fern-mad Americans would want – he separated it from the rest, grew it on and began propagating specimens at such a rate that he was soon selling it by the thousands. Soon Becker had turned almost all his glasshouse space over to the plant he marketed as 'the new Boston drooping fern'. It performed better in homes than the regular sword fern, and within two years, other sellers had begun producing the plant too, and the Boston fern had almost completely supplanted its parent species in terms of sales. As an advertisement by grower W. M. Lott put it in a Pennsylvania newspaper in 1898, 'To see the Boston fern is to want it – to want it is to possess it because the price is so reasonable.' Becker sold his new fern under the scientific name *Nephrolepis davalliodes*, but after some debate and consultation with botanists at Kew Gardens in London, it was settled as 'Bostoniensis', a variety of the sword fern *Nephrolepis exaltata*.

The Boston fern arrived at just the right time: the British had been suffering from fern fever since the 1830s, and Americans were infected too. Nathaniel Bagshawe Ward's invention of the

Wardian case – a glass box resembling a tiny glasshouse – in 1829 opened up a way to give ferns the moisture they needed to thrive, as well as a barrier against the polluted air of people's homes. Fern hunting expeditions became de rigueur, leading to the decline of some species in the wild. Pteridomania was still going strong in the 1890s, so growers were delighted to have a new fern to add to their collections: garden writer Tovah Martin called the Boston fern 'pteridomania's last and loudest hurrah'.

But this was not a fern that could be confined in a Wardian case – it is what modern readers may refer to as a 'chonk'. There are records of healthy Boston ferns reaching 1.5m wide, with fronds well over 1m long: the American magazine *Popular Mechanics* ran a story in 1915 declaring that an unnamed woman from North Carolina had grown a Boston fern with 2.4m long fronds. These are plants that demand respect. American naturalist Gale Lawrence describes going to a friend's house where she met the largest, lushest Boston fern she had ever seen. Appropriately enough, the friend had named it 'Sir'.

Its wild parent, *N. exaltata*, is indigenous to a swath of countries in tropical and subtropical America and the Caribbean, including Florida, Cuba, Jamaica and Brazil; it is also naturalised in various tropical countries including South Africa where it is starting to crowd out indigenous species. In Spanish, sword fern is called *cola de quetzal*, meaning tail of the quetzal, a Central and South American bird with huge tail feathers that bring to mind the fern's fronds. It grows in shade both terrestrially in swamps and moist hollows, and as an epiphyte, lodging itself in various tree species, including oaks, and cabbage palms (*Sabal palmetto*): its fronds have been recorded at lengths up to 8m. Sword fern sap is believed to have insecticidal qualities in traditional medicine, and on the Micronesian island of Pohnpei it is placed around the house to keep cockroaches away: recent scientific research seems to show the plant extract is indeed toxic to cockroaches.

The sword fern was in cultivation long before Becker discovered 'Bostoniensis' – it was sent to Kew Gardens in the 1790s from Jamaica. And ferns themselves are some of the most ancient plants on earth, whose ancestors first appeared 360 million years ago during the Carboniferous period – also appropriately known as 'the age of ferns'. Ferns are true foliage plants, as they do not produce flowers or seeds, opting for a different, more primitive means of sexual reproduction. Mature plants release spores from the under-side of their leaves. When they find a place to settle, the spores grow into tiny plants called prothalli which carry both male and female cells, and it is at this point that sexual reproduction occurs.

Unlike its parent species, the Boston fern is sterile. It relies instead on reproducing itself vegetatively. This is achieved via a rare phenomenon for a fern – fuzzy, leafless stolons (aka runners) that emerge from the base of the plant and grow a new baby plant wherever they hit soil. 'Bostoniensis' was far from the end of the story when it came to the city of Boston's fern habit. This plant is extremely prone to throwing out genetic mutations via the new plants that grow from its stolons, so growers found dozens of muta-tions – known as 'sports' – in the next few decades. Fronds became longer, smaller or split into two; others grew ruffled, or a different shade of green, and soon every grower in the city had their own special fern to market.

The Boston fern became the mainstay of the American foliage plant market through to the first few decades of the twentieth century, particularly in central Florida: as a garden writer noted in the *El Paso Herald* in 1920, 'If a family has only one plant now, it is a Boston fern.' Growers in subtropical climates grew it on the porch year round; those further north grew it inside on plant stands or in hanging baskets. The plant's next heyday was in the 1970s, when houseplants underwent a general renaissance, but Boston ferns had a particularly significant resurgence: according to a 1975 article in *New York* magazine, the oil crisis was pushing up heating bills, and thus 'fern freaks are bound to grow in number', because these plants thrived in cooler rooms. The Boston fern also became a symbol of a new American phenomenon – the 'fern bar'. These were upscale city bars designed to appeal to singles – and specifically women – at a time when the sexual revolution was inspiring women to explore and inhabit traditionally male realms. Forests of Boston ferns hung from the ceilings and fake Tiffany lamps glowed from corners, offer-ing an ambience very different from the macho world of the average American bar at the time.

The fern bar may have faded from memory for generations of young people who are more likely to be found on a dating app than in a singles bar, but the fern lives on in our homes, having been rediscovered by a new generation of growers in the twenty-first century. In 2019, Netflix made a film version of *Between Two Ferns*, a parody of the US chatshow format where real celebrities faced painfully inept questioning by host Zach Galifianakis. Critics wondered whether the ferns were real. Of course they were – aside from the scenes where the actor Matthew McConaughey nearly gets drowned, when they were replaced with plastic 'stunt plants'.

I would not recommend taking Boston ferns on a road trip, however. They are large, unwieldy and much more fragile than, say, a Monstera. But are they an easy houseplant to grow? That really depends. Some people find them a doddle, while others watch one Boston fern after another resolve into a crispy mess within months. I am ashamed to say that I have sometimes fallen into the latter category. American humorist Erma Bombeck summed up the feeling of being a Boston fern failure: 'You have to know that from the moment you take it out of the nursery, it turns hostile. It doesn't want your water. It doesn't want your sun. It doesn't want your fertiliser or your Willie Nelson Records. It just wants to die.' It's easy to feel victimised by a plant with so much presence that refuses to thrive in your home, but try not to take it personally: the people who succeed with Boston ferns are simply those who can fulfil its demands for lots of light, moisture and space.

CARE GUIDE

Light Boston ferns love light: if you can hang yours close to a full-length window, it will be happy, provided it has protection from the rays of the summer sun. They can be moved outside to the shade of a tree or placed in the lee of a wall in summer.

Temperature Although Boston ferns are a popular feature of American porches, they can only be kept outside year round in places where the temperature does not drop below 10°C (50°F).

Water This is a thirsty plant: larger specimens are challenging to water because their rootballs are large and hard to reach through the mass of fronds. In summer, soaking plants in a bucket of water every few days for half an hour then draining and replacing will do much more good than daily trickles of water. Ease back watering in winter but never let the rootball dry out.

Humidity High humidity will help your Boston fern look its best: a combination of high temperatures and low humidity are their worst enemy. Misting is just about the least successful way of increasing humidity around your plant: it's more effective to place plants in a naturally humid room such as a kitchen or bathroom, or stand the pot on a tray of wet pebbles.

Pests and diseases All and any houseplant pests will attack a Boston fern, especially one undergoing drought stress.

Substrate As this species is epiphytic, plenty of air must reach the roots. The RHS suggests growing Boston ferns in one part loam, two parts sharp sand and three parts leaf mould. Given that not all houseplant growers have access to leaf mould, opt for a potting mix that is rich in humus but also airy, with a pH that is neutral to slightly acid: peat-free houseplant potting mix with added vermiculite to improve moisture retention, and perlite to aid drainage. When repotting, ensure that the crown of the plant stays at the surface rather than buried below, as this can cause rot.

Propagation The best way to make more plants is by pulling out stolons from the fern's base: when pinned onto moist soil, these will produce baby plants. If your plant becomes too big, it can be divided: remove it from the pot, slice the rootball into chunks with a sharp clean knife, and pot up separately.

Feeding Use foliage houseplant fertiliser at half strength every couple of weeks from spring to early autumn.

Other maintenance tasks Boston ferns are messy plants: even when they are doing well, they tend to drop the odd frond, so be ready with the dustpan and brush. Damaged or yellowing fronds should be cut away at the base, and some growers suggest winding the stolons around the crown to keep them out of the way. Turn plants regularly to ensure even growth.

Danger signs As soon as they get a little stressed, the response is a potpourri of dead foliage spread below your plant. Plants respond to drought stress by slowing growth and leaves becoming dull and lustreless – this is sometimes known as 'greying'.

Toxicity Non-toxic to pets and humans. Some people find touching this fern causes contact dermatitis.

Display Boston ferns demand a plant stand or hanging basket to show off their arching fronds. Mature plants can be extremely heavy and large, so ensure whatever you use is appropriately sturdy. If you are a fan of *The Simpsons*, why not buy a pot with Sideshow Bob's face on it for your Boston fern and create the ultimate tribute?

Cultivars Many other varieties and subspecies have been discovered since the Boston fern: 'Compacta' is useful if you are short of space, while 'Fluffy Ruffles' has delicately ruffled fronds and 'Rita's Gold' has golden-green fronds.

Also try . . . If you have succeeded with the Boston fern, try the so-called 'macho ferns', *N. biserrata*, a sort of houseplant equivalent of Dwayne 'The Rock' Johnson. My recommendation for those unsuccessful with Boston ferns is the toughest and most drought-tolerant fern I have found: *Davallia tyermanii*, the haresfoot fern. (Arachnophobes, be warned: its hairy rhizomes look like tarantula legs).

Opuntia microdasys

BUNNY EARS CACTUS

Cactaceae

Take some duct tape, apply it carefully to a body part of your choice, then rip it off. If this doesn't strike you as an exercise you'd like to repeat, the bunny ears cactus may not be the plant for you.

The prickly pears – members of the genus Opuntia – have long been of huge economic and cultural significance to the people living in their indigenous range of the continents of North and South America and the Caribbean. From 3000 BCE, Opuntia were being farmed by settled communities. Their pads and fruits are eaten by humans and fed to their animals, used as a source of adhesives, medicines and food additives, and some species are planted as a form of fencing; the cochineal insects (*Dactylopius coccus*, a type of scale insect) that feed on them are processed into a red dye found in many foodstuffs too. *Opuntia microdasys*, though, is close to useless to humans in a practical sense: it is only used as emergency fodder for livestock.

However, *O. microdasys* has gained the status of a popular succulent houseplant, prized for its iconic looks. This plant works on a windowsill because it is one of the more dainty members of the genus; the species name *microdasys* translates as 'small and hairy'. Bunny ears only grows up to 1m tall in nature, and often maxes out at 60cm, growing sideways as much as it grows up: a dwarf compared with the treelike *O. megasperma* indigenous to the Galápagos Islands, and far smaller than many of the other species that grow alongside it in the Chihuahuan Desert of Mexico and the southern US.

The bunny ears cactus is made up of flattened, paddle-shaped stems called cladodes that grow one on top of another in a seemingly precarious pattern that resembles a toddler's attempt at making a tower of plastic bricks. The name bunny ears makes sense when you see a young specimen with a couple of cladodes poking up inquisitively. Each cladode is covered with very fine hairs that are invisible to the naked eye, but offer a velvety look; they are also dotted with yellow spots, which may look as plush as a fluffy bunny, but contain the plant's secret weapon: glochids. Each dot is an areole: a bump that acts as an anchor for around 100 glochids. These tiny hairlike structures made from cellulose may look harmless, but stick one under an electron microscope and their purpose becomes clear.

Each glochid measures around 1–3mm long, with a pointy tip. Down the sides of the top third or so of each glochid are minuscule barbs. Although glochids detach from the areole with ease, once they are lodged in the flesh, their barbs act like hooks and make them hard to remove, causing much irritation, and conjunctivitis in eyes. This damage isn't limited to humans, either. In Mexico *O. microdasys* is called *nopal cegador* – blind prickly pear – because the glochids have been known to hurt the eyes of grazing livestock. So, if you own a bunny ears, take care to ensure they cannot be reached by children, pets or any unsuspecting person, and handle them only when wearing extremely thick leather or rubber gloves; surround the plants with a thick layer of rolled-up newspaper as an extra precaution when repotting.

Even then, glochids may still find a way to get you. Spend time in the company of cactus collectors and the subject of the optimum way to remove glochids will inevitably come up. If you do get spiked, do not touch your face or try to suck out the glochids with your mouth: this will just spread the pain. One of the most commonly mentioned cures involves applying duct tape or a thick sticking plaster (adhesive bandage) to the affected area and ripping it off, along with individually removing glochids using tweezers, with a bright light and magnifying glass to hand. Washing the area and brushing with a stiff brush are also recommended in the medical literature. According to the *American Journal of Emergency Medicine*, glochids are best removed by applying a layer of professional facial gel, allowing this to dry, and then peeling this away. Hair removal mitts were recommended in another medical case study. Having scared you to death about the perils of glochids, I should mention there is a way to grow a bunny ears that will not harm you. 'Angel Wings' is a variety thought to have come about as a natural mutation of the species, first appearing for sale in 1949, and is reputed to be one of the best-selling cacti in existence. It has white glochids which are unbarbed and therefore harmless. Do double check before you buy, as angel's wings is also a common name applied to *O. microdasys* more generally.

Of course, glochids have not evolved to wind up cactus growers: they protect the plant from being munched by herbivores, which have learned not to snack on them. The glochids also assist with another water-saving function. When the soil is dry, *O. microdasys* has another way of finding the moisture it needs: collecting water droplets from fog. Using time lapse photography of a single glochid, a group of Chinese scientists led by Jie Ju and Hao Bai documented the way that bunny ears collects water. They found that the barbs at the top of the glochid capture tiny droplets of water, which are then channelled down towards the plant via grooves in the surface, regardless of whether the hair is pointing up or down. The water is absorbed by clusters of belt-like trichomes or hairs that sit at the base of the glochids. By replicating these complex structures, scientists are creating larger scale fog collectors to help supply water in arid environments.

That ability to stay hydrated by collecting fog means bunny ears can perform its role as a valuable part of the ecosystem, providing shelter for birds, insects and small mammals, as well as pollen, nectar and fruit for invertebrates. The home of the bunny ears cactus is the Chihuahuan Desert, the largest desert in North America, with four fifths lying in Mexico and the other fifth in the south-western US – although bunny ears originates from Mexico. Although this desert lacks huge columnar cacti such as the saguaro (*Carnegiea gigantea*) that characterise other cacti-rich areas, it's still extremely rich in plantlife; depending on who is counting, it contains a third or a quarter of the world's 1,500 or so cactus species. Unfortunately, *O. microdasys* has become an invasive species in other parts of the world, creating dense thickets that can crowd out local species. One of the explanations for this spread is that people unknowingly buy one as a houseplant, have an unfortunate tussle with its glochids and end up throwing it out. Couple the first class vegetative propagation skills of this species – it can grow a new plant from a single pad or part of a pad – and it's not hard to see how the plant has, like other members of the genus, become a problem.

The reddish-purple fruits that follow on from the flowers can also help the plant spread, in one obvious and one surprising way. The obvious one first; birds eat the fruits, fly off and then defecate the seeds onto the ground. Sometimes, though, the fruit fails to ripen and drops off the plant. Rather than rotting away, the fruit is able to root and become a new plant, a phenomenon known as pseudovivipary. The reasons for this aren't fully understood, but by aborting fruit production at an early stage when conditions aren't right, the plant has maximised its potential to stay alive by allowing fruits to have a second function as genetic clones of the parent plant.

The species was first described by European botanists in 1827, and caught on quickly in cultivation, despite its drawbacks: it was not only the glochid issue, but the fact that, when grown as a houseplant in a pot, *O. microdasys* is unlikely to flower, as it needs open ground to stimulate development of the lemon yellow flowers. Nevertheless, by 1847 *The American Flower Garden Companion* noted that grafting various epiphyllums onto *O. microdasys* was 'becoming general among amateurs'. Grafting is a technique for joining two often unrelated plants by means of cutting them and joining them together until the tissues fuse; the botanical equivalent of a 'cut and shut' car. I can't imagine how many glochids lodged in the hands of these Victorian amateur grafters, but I salute their efforts. The bunny ears cactus features in many twentieth century seed and plant catalogues, although of course the nurseries always emphasise its beauty and neglect to mention its unpleasant aspect. By 1867, *Wonders of the Vegetable World* by William Davenport reported that this 'robust thorny plant of fantastic appearance' is 'now common in English conservatories'.

CARE GUIDE

Light As much sunlight as possible is required for the bunny ears cactus to thrive. You can place plants outside in summer, but increase their exposure to the sun outside gradually, as it is much harsher than the light indoors. Plants that are lacking in light will produce thin, finger-like cladodes rather than rounded ones.

Temperature This species can withstand occasional light frosts, and will thrive in an unheated glasshouse in temperate climates where light frosts are the norm over winter, provided it is kept completely dry at this time. Indoors, a cool winter rest is advisable, so you could move plants to a cooler room where there is no heating such as a porch or spare bedroom, but make sure there is still maximum light; boost with a growlight if necessary.

Water Provided the substrate is free draining (see below), bunny ears can be watered generously from May to September; be sure to remove any excess water that gathers below the pot once the substrate has been fully wetted, then water again once the rootball is dry.

Humidity Dry air is not a problem for this cactus.

Pests and diseases Mealybugs and scale are the usual suspects.

Substrate A free-draining mix is essential: every cactus collector has their own recipe, but half to two thirds houseplant potting mix or John Innes No. 2 combined with some kind of drainage material works well: that could be perlite, grit or horticultural sand.

Propagation Twist off a pad (making sure to protect your hands with thick gloves or a wad of newspaper) and leave for several days so the wound can dry out, then place on gritty substrate.

Feeding Use a specialist cactus feed as per the instructions during the growing season.

Other maintenance tasks If your plant becomes dusty or picks up some particles of potting mix during repotting, carefully brush away with a soft, clean paintbrush or makeup brush. Spray with water occasionally when in active growth to remove dust.

Danger signs Round brown spots on the cladodes can be caused by sunburn, or by cold temperatures. Check also for root rot, which is caused by too much water around the roots, especially in winter. Plants that start to droop may be too dry, but also check for pest infestations and root rot.

Toxicity No known toxicity to humans or animals, but avoid touching the glochids.

Display Whatever pot you choose, it must have drainage holes: one large, or several smaller ones. Do not torture your Opuntia by thinking it will love life in a glass terrarium. I prefer terracotta pots to plastic ones, as they allow more evaporation of excess moisture from around the roots, but they are also more prone to damage. Glazed pots work well provided they have drainage holes, and both terracotta and glazed pots are usefully heavy and prevent the plant from toppling over. *Opuntia microdasys* roots are shallow, so a shallow dish-style pot works well.

Cultivars Look out for cultivars of *O. microdasys* with pure white, rusty red or brown glochids; if you have an eye for the unusual, the cristate form, *O. microdasys* 'Cristata', has attractively wavy cladodes.

Also try . . . There are many other relatives of the bunny ears cactus worth growing as houseplants: my favourites include *O. monacantha*, which is often sold in a variegated monstrose (twisted) form, and *Austrocylindropuntia subulata*, commonly known as Eve's pin, after its curious needle-like leaves.

Oxalis triangularis

FALSE SHAMROCK

Oxalidaceae

Chewing on a leaf of this plant is akin to sucking a mouthful of sherbet lemons. Your lips will purse, one eyelid may quiver alarmingly, but you will feel truly alive. *Oxalis triangularis* falls into a small subset of houseplants that are palatable and non-toxic to humans, but you'll need to grow a lot of plants to guarantee a supply of leaves for anything more than a garnish on a salad or a cocktail. (See edibility note below.)

Oxalis triangularis is a member of the wood sorrel family (Oxalidaceae), and a relative of oca (*O. tuberosa*), a staple food of Andean South America prized for its edible tubers. Some wood sorrels have spread way beyond their home range to become world beating weeds: most notably *O. corniculata*, which spreads across the surface of nursery pots, perhaps even invading the odd houseplant pot. The name for the genus, Oxalis, comes from the Greek word *oxis*, meaning acid: the wood sorrels taste sour because they are packed with oxalic acid. Scientists don't fully understand why our faces pucker when eating anything as sour as an Oxalis leaf, but it's thought to serve a couple of purposes. First, it signals that whatever we're eating could be bad for us; unripe or rancid fruit, for instance. Second, tightening the muscles of the face channels saliva onto the tongue, which helps to stop the mouth's pH levels dropping so low that the acid damages tooth enamel. These saliva-inducing qualities explain why wood sorrel leaves have historically been eaten as thirst quenchers. *Oxalis triangularis* has also been found to have significant antibacterial qualities, and can knock out major bacteria including Staphylococcus, e coli and salmonella.

Sales of this plant often peak around St Patrick's Day on 17 March, but *O. triangularis*'s links to Ireland do not extend beyond its vague resemblance to the true shamrocks (*Trifolium dubium* and *T. repens*). To find false shamrock in the wild, we must head to South America, where they proliferate from Peru, through Brazil and to northern Argentina. The species has also become naturalised in two US states with similar climates, Louisiana and Florida. It proliferates in wet places: along the sides of streams and in marshes – something that's worth bearing in mind when you grow it indoors. If you find this plant growing in South America, the chances are it will have green leaves. And yet the most commonly available false shamrock sold as a houseplant has purple foliage: taxonomists are still discussing whether this is a subspecies, or simply a variation of the species, and you may still find it referred to by an outdated scientific name, *O. regnelli*. For now we'll call it *O. triangularis* subsp. *papilionacea*. It may feel as if it's been around forever, but you won't find this species in houseplant books that pre-date the 1990s, as it was only described to science in the early 1980s.

If your style of houseplant care can be categorised as erratic, this is one to try, because it is a plant with a backup plan. False shamrock grows from scaly rhizomes – a kind of modified underground stem – that look a little like a bunch of underfed pinecones; or as one particularly imaginative garden writer put it, 'like lawn grubs having an orgy'. Those rhizomes give the plant the ability to survive inclement conditions: both drought and low temperatures. The foliage will die back, but below ground, the rhizomes store water and nutrients, ready to burst forth new leaves when things improve. Unearth older specimens and you may find the rhizomes are connected to a juicy taproot that looks like a white daikon radish and is also edible.

You may spot false shamrock growing outside in subtropical or even temperate climates as a herbaceous perennial that dies back in winter, or as temporary summer bedding. When kept as a houseplant, *O. triangularis* also has a habit of going dormant as the days shorten, dropping some or all of its leaves. I see this as a bonus rather than a drawback, meaning you can allow the whole pot to dry out, store it somewhere for a couple of months, then start watering again in spring and your plant will re-emerge. If you don't have space to store the whole pot, carefully dig out the rhizomes, lay them on a plate for a couple of days to dry out, then store in a paper bag somewhere cool and dry until you are ready to replant.

The subspecies has dark purple leaflets, with a lighter splodge in the centre. The red pigments in these leaves – known as anthocyanins – act like sunscreen, protecting the bodies called chloroplasts that exist in every cell and carry out the process of turning sunlight into energy. They are a powerful pigment, so expect to stain your hands if you accidentally crush a leaf. This explains why the plant has been investigated as a natural replacement for synthetic dyes.

Oxalis has the charming habit of closing up its leaves at night, folding them in neatly like a parasol. This movement is known as foliar nyctinasty and is a fairly common phenomenon in the plant kingdom, seen in everything from the tiniest plant to the largest tree. You may never have noticed that peas and beans also close up their leaves at night, for instance. But nyctinasty among houseplants is more obvious, because they live alongside us. There are theories galore attempting to explain foliar nyctinasty: protecting leaves from frost damage; allowing foliage to shed water; and protecting them from being munched by herbivores. A more recent theory posited by biologist Peter V. Minorsky suggests that closing up of leaves assists the predators of nighttime herbivores, and fewer herbivores equals reduced leaf damage. This theory considers the plant, the herbivore and its natural enemies as an interdependent network. How does this work? Theories include the idea that reducing the leaf surface from above allows predators such as bats and owls a greater chance to spot and kill leaf munchers by allowing more light to reach their potential prey; less surface area of leaves also means less acoustic dampening of the signals of bats and owls as they hunt, improving their kill rate; volatile chemicals produced by herbivores that signal their presence to predators are disrupted by foliage, but less so when leaves are folded; finally, herbivores are less likely to forage where there is reduced foliage cover.

False shamrock foliage has another eye-popping ability, known as superhydrophobicity, or the lotus effect. You'll notice that when you mist a plant, the leaves shuck off the water in droplets and are not easily wetted: the poster leaf for this particular planty talent is the lotus (Nelumbo species). These have a microscopically bumpy, waxy surface that holds a tiny layer of air around it. When a water droplet lands on a leaf, it sits as if on a microscopic bed of nails, with just 2–3% of its surface in contact with the leaf, and subsequently rolls off with ease. This helps keeps leaves clean, as particles of dirt, fungal spores or algae will be picked up by the droplets as they slip away. Scientists studying superhydrophobicity in *O. triangularis* have found that the leaves are covered in microscopically small, convex star shapes, with grooves in between that hold air and reduce contact between droplets and the leaf. This kind of research offers up possibilities of mimicking nature to create superhydrophobic surfaces to fill specific human needs, such as self-cleaning buildings and non-fogging glass, and even surgical facemasks with superhydrophobic surfaces to repel saliva and other bodily fluids.

One of the other common names of *O. triangularis* is the love plant. I haven't managed to find out why, except for the triangular leaves' resemblance to a heart shape. But the more I discovered about its incredible properties – edible, foldable, antibacterial, water repellent – the more I came to agree that it had earned this moniker.

CARE GUIDE

Light The colour of the foliage will fade in direct sun. Other than that, it is adaptable to most conditions, but will grow more and bigger leaves the more light it gets. So if your Oxalis is small, move it closer to a window.

Temperature It will tolerate minimum temperatures of around 12°C (54°F). Less than that and the plant may die back, but unless the rhizomes below ground are frozen, it should regrow.

Water Given its presence alongside streams in the wild, it should be no surprise that false shamrock likes moist soil: its water-storing rhizomes allow it to cope with the substrate drying out, but your plant will grow far more lush if given a steady supply of moisture around the roots. As well as closing up at night, it will also fold its leaves during the daytime in particularly hot dry weather, especially when the soil is also dry, so take this as a sign that it needs checking straight away: give it a good soak in the shade and it should quickly revive.

Humidity Regular humidity of 40–50% will be fine.

Pests and diseases Spider mites can infest plants if conditions are too hot and dry: watch out for the plant looking generally miserable, and for white grainy deposits on the undersides of the leaves, which will also display a pale stippling where the mites have sucked the sap: a harsh but effective treatment is to simply cut off all the leaves. Foliage pockmarked with bright yellow circles are a sign of oxalis rust, caused by the fungus *Puccinia oxalidis*. Pick off and dispose of affected leaves immediately (remember you can take off every single leaf and the plant will regrow). Spray fungicides can be used if necessary.

Substrate Use regular houseplant potting mix; if you are prone to underwatering, choose a self-watering pot or add a little vermiculite to the potting mix to aid water retention.

Propagation This couldn't be easier: separate out a rhizome or two and pot up separately to make a new plant. You can even put an individual leaf with its stem but no rhizome attached into a glass of water, and it will root, although this is a slower procedure.

Feeding Use a foliage houseplant feed regularly when the plant is actively growing.

Other maintenance tasks
The five-petalled flowers are, for me at least, the least exciting part of the plant. I remove the white or pale pink blooms from the plant with a sharp yank at the base of the spindly stem as they appear, preferring instead for the plant to put its energies into producing foliage. If you disagree, leave the flowers be.

Danger signs Large numbers of leaves starting to die off is often a sign the plant is going into dormancy in winter; the rest of the year, it can be due to drought or waterlogging. Plants will become sparse in low light conditions.

Toxicity Toxic to pets but edible in moderation for humans. False shamrock's tangy flavour comes from the presence of oxalic acid, the same substance found in other foods including rhubarb, spinach and endive. It's wise to eat oxalic acid-rich foods in moderation as it can hamper calcium absorption, particularly if you suffer from gout, rheumatism or kidney stones.

Display A forest of false shamrock can look wonderful as ground cover under really big specimen plants, although it may die back in winter. I place pots on the floor so I can enjoy the leaves from above.

Cultivars More cultivars are becoming available, offering bigger or smaller leaves and different colours. Both 'Fanny' and 'Irish Mist' have green leaves speckled with silver.

Also try... *Oxalis corymbosa aureo-reticulata* has green leaves with a fine network of yellow tracing the veins, while *O. vulcanicola* has smaller leaves and yellow flowers.

Pilea peperomioides

CHINESE MONEY PLANT

Urticaceae

*P*ilea peperomioides* is a curious anomaly in the houseplant world: a species forgotten by botanists for more than 60 years while proliferating in homes across Scandinavia and the UK.

When Scottish plant collector George Forrest found his first specimens of *P. peperomioides* during a plant hunting trip to western Yunnan, China, in 1906, he was returning to the scene of a drama that nearly killed him. His exploits earned him the title of 'the Indiana Jones of the plant world' in modern accounts, although these retellings gloss over the many indigenous people who lost their lives helping Forrest track down plants for European gardeners.

In 1910 the *Gardeners' Chronicle* published an article by Forrest entitled 'The Perils of Plant Collecting'. He notes in the introduction that the constant threat of diseases and wild animals are among those perils, but adds: 'Not infrequently too, the collector has to seek his specimens among savage or semi-civilised peoples who, in most instances, strongly resent his intrusion into their midst.' Forrest certainly experienced 'resentment' during his trip to Yunnan in 1905. The British had invaded the region in December 1903, and in March the following year turned their machine guns on 1,500 Tibetans armed only with muskets and broadswords: around 700 of them died. By the time Forrest arrived in Cigu (then known as Tzekou), the Tibetans had begun a rebellion, and he and his retinue were in extreme danger. He fled from the French Catholic Mission where he had been sheltering along with the French missionaries and their followers as the Tibetans approached. In the ensuing scramble to escape, 66 out of 80 people in the party died: just one of the 17 locals Forrest had employed as collectors and servants survived.

Although Forrest lost almost all of the samples he had collected on this trip in his rush to escape being hacked to death, he was – incredibly – not put off from returning. His visits in 1906 and 1910 produced thousands of plant samples and pounds of seed that were brought back to Britain, including two samples of *P. peperomioides*. While some of the plants Forrest collected became popularly cultivated plants, including the candelabra primrose (*Primula bulleyana*), and the diminutive yellow orchid *Pleione forrestii*, the Chinese money plant does not seem to have made an impact at the time. Plant samples from both visits were deposited in the herbarium at Edinburgh's Royal Botanic Garden, and given the scientific name *P. peperomioides* by German botanist Freidrick Diels in 1912. There they stayed, pressed between the pages, for several decades, until 1978 when a botanist from the Royal Botanic Gardens in Kew called.

Earlier that decade, Kew began to receive requests for help identifying a houseplant with round leaves. Without any flowers to examine – flower structures allow botanists to delineate one species from another – they could only speculate as to what they were, beyond a general impression that they were part of the nettle (Urticaceae) or pepper (Piperaceae) family. Then in 1978, a Mrs D. Walport of Northolt, west London, sent a specimen with flowers attached. After being passed around various botanists, South African Wessel Marais identified it with the help of a loan of Edinburgh's herbarium specimens collected by Forrest. That did not solve the mystery of how the plant had ended up in Northolt, however. In 1983, *Sunday Telegraph* columnist Robert Pearson joined the cause, writing an article asking readers to help solve the mystery. Step forward Jill Sidebottom, a nine-year-old girl from St Mawes in Cornwall, whose parents reported that they had brought back a plant after visiting Jill's former au pair Modil Wiig in Norway.

The next step was to discover how the plant made it from China to Scandinavia. Botanist Lars Kers of Bergius Botanic Gardens in Stockholm spotted the *Sunday Telegraph* article and an article in Kew's magazine, and knew he could finally put a name to the plant he had owned since 1976. Kers appealed for information on Swedish TV in 1984 – 10,000 letters poured in from people happy to report that they grew this plant. One of the letters was from Knut Yngvar Espegren, the son of a Norwegian missionary to China who had fled from his posting in Hunan in 1944 due to the precarious political situation there. Agnar Espegren and his wife and four children had been flown out in an American plane, and during a stopoff in Kunming, Yunnan, he picked up a plant, boxed it up and headed back to Norway, via a stopoff of some months in Calcutta. Kunming is not that far from the Cang mountain range where Forrest collected his Pilea specimens, so while we do not know exactly where Espegren's plant came from, it seems close enough to have been perhaps growing wild, or was maybe sold in a market.

The plant survived and Espegren began to give its offspring to friends and family in Norway, earning it the common name *missionärsplanten* or missionary plant. The plant spread steadily throughout Scandinavia, making it to the UK sometime in the 1960s via Jill Sidebottom and perhaps other British visitors to Scandinavia. When I interviewed Phillip Cribb, a botanist at Kew at the time, for my podcast *On The Ledge*, he told me that some people who brought the plant to Kew for identification told him that their source was the flamboyant British hairdresser and celebrity of the 1950s and 60s, Raymond Bessone, aka Mr Teasy Weasy. I've ploughed through several archive videos of this elaborately coiffed hairdresser for signs of a Pilea leaf poking from a corner, but no such luck.

So, by the mid-1980s, the mystery plant had a name, but it did not result in a huge boost of interest. Dr David Hessayon's 1987 book *The Gold Plated Houseplant Expert* makes no mention of it, and it only starts to appear in houseplant books dating from the 1990s onwards. It took another 30 years before the wheel of fortune turned again for this species. Just as I began making my podcast in 2017, the Chinese money plant became the desirable houseplant of the moment. Facebook groups were set up devoted to helping people desperate for a plant to track one down. Plants were changing hands for $50 in the US: five years on, and it was available for a few pounds or dollars, and is sold by the million at stores like IKEA.

What is it that has proved so captivating about the Chinese money plant? Like several of the species profiled in this book, it is a prolific parent, and once mature it will begin an endless succession of plantlets that form via the basal shoots of the plant and along the central woody stem as it grows. It is not unusual to find plantlets growing out of the drainage holes in the base of the pot. Thus it could spread easily; growers plucked out and potted on a plantlet or two to give to friends and neighbours, earning the species an array of common names including friendship plant and pass-it-on plant.

Its appeal in the twenty-first century is visual too: the coin-shaped leaves attached to stringy petioles arranged around a woody stem look arresting within the square of an Instagram post. Botanists call this leaf shape peltate: round, with the stalk attached to the leaf's underside rather than its edge. It is this shape that inspired the Chinese common names for the plant (the name Chinese money plant seems to have been applied by botanists at Kew after it was correctly identified there): it is called *Jìngmiàn cǎo* – mirror grass – or *Jīn xiàn cǎo* – golden thread grass. The first of those names relates to round bronze mirrors used in ancient China that often featured a nub in the centre to allow attachment to clothing, echoing the spot in the middle of a Pilea leaf where it joins the stem. Golden thread grass refers to the colour of this dot.

The plant is usually described by botanists as either succulent or semi-succulent, and subsequent research has found it can perform the special form of photosynthesis known as CAM: the first species in the nettle family to be found with this photosynthetic capacity (more on this in the *Crassula ovata* chapter, page 45). New growers are sometimes alarmed by the tiny white dots present across the leaf blade, especially on the lower surface, fearing mealybugs or powdery mildew. These are tiny deposits of calcium carbonate known as cystoliths, which sit in specially adapted cells called lithocysts. It is not completely clear what purpose cystoliths serve, although it is thought that they help to scatter light to the centre of leaves to optimise photosynthesis: useful for a plant that grows in relatively shady positions.

The species does not seem to have been widespread in the wild in Forrest's time, as it was not collected by other botanists visiting the area, and today it is classed as likely to be endangered. There is much houseplant growers can learn from considering the settings where this plant grows in the wild. The specimens collected by Forrest were found growing as lithophytes: plants that grow on rocks coated with just enough organic material to give them nutrients, moisture and a place to anchor. The climate can produce short periods of subzero temperatures in winter: reports as to the minimum temperatures the plant can withstand range from 4°C (39°F) to −9°C (16°F), so this may be a plant to grow outside in sheltered spots in temperate climates, or in unheated or cool conservatories and brightly lit rooms; the foliage may be damaged by frosts, but it will regrow once spring arrives. The Chinese money plant can cope with a wide range of light conditions, and will even take a sunny windowsill if given time to acclimatise.

One outcome of giving your *P. peperomioides* a cool room and lots of bright light is that it may flower. Specimens grown as houseplants tend to bloom infrequently, as most people keep them in centrally heated rooms. Given their propensity for vegetative propagation and the fact that the flowers are unremarkable, it is not a great loss, either in terms of propagation or in terms of beauty. The plant is monoecious, meaning it produces pinky-white male and female inflorescences (clusters of flowers) on the same plant. But it often fails to produce the female ones. The male flowers frequently drop off without releasing any pollen, although a spray of water or a gentle squeeze has been found to trigger opening, and then puffs of pollen are sent out. Given that *P. peperomioides* often exists close to water in the wild, is this a vital part of its sexual reproduction process? We don't know - yet. With millions of plants in circulation now, perhaps the answers will be revealed via citizen research by eager growers.

CARE GUIDE

Light This is not a plant for that darkest corner of a room. Some growers have acclimatised their specimens for full sun; most seem to find a sweet spot where their plant receives morning sun in an east-facing window, or sits close to a large north-facing window. Leaves will gain a reddish tinge when exposed to direct sun.

Temperature Warm, stuffy, centrally heated rooms are to be avoided. It can cope with temperatures to and below freezing for short periods, so you can place it in an unheated room, but it may shed leaves.

Water Succulent leaves mean that *P. peperomioides* can cope with periods of dryness, but it looks its best when watered whenever the top half of the substrate is dry.

Humidity This plant comes from a humid environment, so try to keep it at least 50%.

Pest and diseases Mealybugs, aphids and scale are the most likely suspects.

Substrate Look for a quick-draining potting mix: over-the-counter cactus and succulent substrate is one possibility, or use regular houseplant substrate but add a handful of perlite, pumice or other drainage material.

Propagation Plants will start to grow plantlets once they are mature. Individual leaves can also be rooted, provided you cut them from the plant with a small piece of stem rather than trying to root a fallen leaf.

Feeding Feed regularly with a feed for foliage houseplants when the plant is in active growth.

Other maintenance tasks Rotating the plant a half turn every couple of weeks will help the main stem grow vertically. The leaves benefit from a regular wash to remove dust and debris.

Danger signs If a large number of leaves turn yellow and fall in swift succession, check for root rot. Plants suddenly exposed to sun will develop straw-coloured burned patches. The leaves often become somewhat distorted or curled when conditions (light, humidity, water, nutrients) are not quite right. This doesn't signal imminent death, but may signal the need to tweak your care regime.

Toxicity No known toxicity.

Display Once plants get lanky, in nature they will naturally lean until they touch the ground; it's fairly normal for the lowest leaves to turn yellow and fall off, leaving the woody stem on display. If you prefer to let your plant stay upright, it can be staked. Alternatively, behead the plant and root the top cutting to make a new, shorter plant; the base of the plant will also resprout. If you lack space for a big plant, raise a steady supply of young plants to replace the parent plant once it has outgrown your space. Terracotta, plastic or glazed china pots work, but if you use the former, you will need to water more often to account for the water lost from the potting mix by evaporation.

Cultivars A couple of cream-splashed cultivars have come onto the market in recent years, including 'Mojito' and 'Sugar'.

Also try . . . The genus Pilea contains several houseplants that are old favourites, including the aluminium plant, *P. cadierei*, and the artillery plant, *P. microphylla*. The Chinese money plant is often confused with two other similar-looking houseplants: *Peperomia polybotrya*, which has thicker leaves and stems, and *Begonia conchifolia*, which has slightly more pointed peltate leaves and a red dot where the petiole joins the leaf underside.

Sansevieria trifasciata

SNAKE PLANT

Asparagaceae

In the introduction to her book on snake plants, American nurserywoman Hermine Stover lays out a series of instructions for replicating her mother Mildred's so-called 'sansevieria ashtray'. First, find yourself a small, dim apartment in Brooklyn, New York, and a glazed pottery container with no drainage holes. Then dig up some soil from the nearest car park, adding lumps of concrete and asphalt. Add the plant, firm the soil until it is rock hard and provide a trickle of water every two months.

Tongue in cheek this may be, but Stover is right: this is a plant so resistant to neglect that it will survive all manner of mistreatment, including being used as an ashtray. Stover is by no means the first writer to poke fun at the seemingly indestructible qualities of the snake plant. The most famous houseplant writer of them all, Dr David Hessayon, remarked in *The Gold Plated Houseplant Expert*, published in 1987, 'If all else fails, grow Sansevieria.' It is frequently likened to another plant with a cast-iron constitution, the Aspidistra (see page 13), although there is one major difference: Sansevieria is a tropical plant that will not tolerate hard freezes in the same way. If you are going to kill your snake plant, a prolonged excess of water around the roots and a freezing cold room will finish it off.

The snake plant's reputation as indestructible has led to its dismissal in some houseplant literature as a pedestrian, workaday plant, fit only for shoving in a dark corner because, unlike so many other species, it won't die there, or at least it will hang on for a good while. It does not help that *Sansevieria trifasciata* is burdened with an English common name, mother-in-law's tongue, that is unfortunately still common parlance despite its reference to an outdated and misogynistic stereotype. If you are looking for alternative common names, there are plenty of better ones. The Ibo people of Nigeria call this plant *èbùbè agú*, which means majesty of the leopard, and in Yoruba it is *ọ̀já kòríkò* (hyena's girdle) or simply *pàçánkòríkò* (hyena). In Brazil, the plant is known as *espada de São Jorge* (sword of St George). And I learned from Ulrich Haage, owner of the German cactus nursery Kakteen Haage, that in East Germany before the fall of the Berlin Wall, Sansevieria was known as *die SED blume*, or the SED flower – SED being the governing Socialist Unity Party of Germany – because this plant was a guaranteed presence in government offices and buildings.

In 2017, taxonomists concluded that the Sansevierias should be incorporated into the genus Dracaena, although the world of house-plants has yet to catch up, and the new genus allocation is still being hotly debated by botanists and horticulturists: I am sticking with Sansevieria for the moment. *Sansevieria trifasciata*'s home range is seven countries in West Africa, including Nigeria, the Democratic Republic of Congo and Equatorial Guinea. It has practical, spiritual and medicinal applications: the leaves are absolutely packed with fibres made from cellulose and lignin, which give the plant its stiff, upright appearance. Once removed and dried, this fibre is strong enough to be used to make everything from baskets to bowstrings, and has also earned it the name bowstring hemp. In the African countries of its origin, in other places where it has been grown and naturalised, and in the African diaspora in Brazil in particular, *S. trifasciata* is used as a protective charm against the evil eye and as a defence against bewitchment, and is an ingredient in several Afro-Brazilian religious ceremonies. Traditional healers employ the leaves for various purposes too: as a treatment for snake bites in Uganda and Bangladesh, for instance.

Sansevieria trifasciata isn't the most dramatic of the snake plants: its sword-like leaves are dark green, etched irregularly with horizontal bands of silver, and it is reluctant to flower when grown in a pot. *Trifasciata* simply means in three bundles, which refers to its flowers, or possibly the silvery bands on the leaves, depending on which botanist you ask. With leaves that reach up to 1.5m long, it is not the biggest species, either – that title probably goes to the elephant tusk plant, *S. stuckyi*, whose sharply pointed, cylindrical leaves can reach more than 2m tall in habitat. However, I chose to include *S. trifasciata* as it is the species that has been most successful at colonising the world's homes through its many cultivars and varieties. The most recognisable and popular is the variety 'Laurentii', with banana-yellow stripes running up the margins of the leaves. Its origins are unknown, other than that it came from the Democratic Republic of Congo, and was found growing near what was then Stanleyville (now Kisangani) by Belgian botanist Emile Laurent: it is assumed it arose from a natural mutation rather than anyone's breeding efforts. Nicholas E. Brown in his 1915 monograph calls this 'A rather remarkable plant', reporting that efforts to reproduce the plant via leaf cuttings ended in failure, in that the yellow edges were not reproduced in the offspring. 'Laurentii' certainly outstrips its parent in terms of popularity; the plainer dark and light green leaves of the species are less often seen in houseplant shops than the more colourful varieties. *Sansevieria trifasciata* does not bloom readily when grown in a pot, but mature plants may reward you with a spike of creamy, scented tubular flowers.

The plant's potential as a fibre crop caught the attention of the US in the 1940s and 50s, when global instability prompted government researchers to plant acres of Sansevieria in Florida in an attempt to develop a homegrown alternative to fibre imports such as sisal (*Agave sisalana*), abaca (*Musa textilis*) and henequen (*Agave fourcroydes*). Various species were trialled, but *S. trifasciata* was found to produce the best crop. However, difficulties in processing the leaves, and the arrival of man-made fibres such as nylon and rayon in the 1950s, put paid to further work in this area.

At the same time, as a decorative plant across the globe *S. trifasciata*'s star was rising; outside in tropical climates, and inside elsewhere. Snake plants reached a peak of popularity in the 1950s, prized for their minimalist form which echoed the sleek lines of mid-century design. As British newspaper the *Yorkshire Post* noted in 1954, it is a plant which 'goes with contemporary furniture in much the same way that aspidistras go with antimacassars'. In 1956, English china firm Midwinter issued a new pattern called Plant Life, by designer Sir Terence Conran, featuring a snake plant arrayed in a plant stand. And set designers on the *Mad Men* television show set in the 1950s filled a trough planter in main character Don Draper's swish apartment with *S. trifasciata* 'Laurentii'. The snake plant began to lose ground by the 1960s: in 1956, it made up 16%

of the total foliage plant mix produced in America's major house-plant producing state of Florida; by 1975 this figure had dropped to 3%.

When Hermine Stover's book on Sansevieria came out in 1983, the plant was still largely ignored. Dedicated collectors continued to enjoy their snake plants, but they seemed of little interest to anyone outside that rarified world. Until the 2010s, many Dutch nurseries had stopped cultivating the many cultivars and varieties of snake plants, as there was just so little demand. It took a few decades, but the rehabilitation of Sansevieria is finally under way. By 2017, those same nurseries were scrambling to locate mother plants of the key Sansevieria species and cultivars to begin mass propagation in the hope of meeting rapidly growing demand. By 2020, Sansevieria was established as one of the top 10 most searched-for houseplants online.

The story wasn't over for the snake plant when it comes to its use as a fibre plant, either. As the climate crisis forces manufacturers to seek out plant-based alternatives for materials, Sansevieria fibre is rising in importance once again: researchers are looking at how it can be used as an ingredient in composite polymers in the aerospace industry. Domestic products utilising Sansevieria fibres are on the rise too: there are artisan makers producing handbags and hammocks, while one company has created biodegradable sanitary towels made from the fibres coated with extracts from the damask rose (*Rosa* x *damascena*). Meanwhile, preppers, foragers and survivalists are taking to YouTube to show how to create cordage from snake plants by separating the fibres from the leaves using a credit card, although the traditional method is water retting, where leaves are steeped in water until the flesh rots away, leaving the fibres behind. Scientists are also exploring how to harness its other qualities. They have discovered that *S. trifasciata* is efficient at removing cadmium from soils, which could be a useful, environmentally friendly tool for dealing with cleanups of ground polluted with this heavy metal.

CARE GUIDE

Light Grown as a houseplant, it's going to be almost impossible to give a snake plant too much light. It is completely possible, however, to give it too little. They tend to be slow growers, but will completely stall if put in a dark corner. They are often recommended for low light, but there are better plants for these settings than Sansevieria. Take care not to suddenly increase light levels, as this can burn leaves.

Temperature Sansevieria can survive down to 7°C (45°F) or lower, but will need a porous substrate that is kept completely dry in those conditions. Do not let them freeze. An extended period of strong light and high temperatures over summer is reputed to encourage flowering, but this also depends on the species: *S. trifasciata* flowers infrequently.

Water The leathery leaves will put up with months of dryness but can be killed by poorly drained soil that prevents air reaching the roots. Regular soakings during summer should be given once the substrate is dry around the rootball. Slow down watering in winter so the plant can rest.

Humidity Dry air is not a problem. Moist air plus very low temperatures can cause problems in terms of rot, so if your plants are in a cool greenhouse, ensure adequate ventilation.

Pests and diseases Not usually a problem in healthy plants, although mealybugs can show up.

Substrate Snake plants are amenable to growing in hydroponic systems, or in water, where their orange roots look rather attractive through the glass. But if you choose an organic substrate, use one for cacti and succulents, or make your own mix from two thirds John Innes No. 2 or similar loam-based substrate, and one part drainage material, such as perlite or grit.

Propagation Division is best: leaf cuttings will work, but it is slower and variegation will not show up in new plants grown this way, and they will revert to the look of the species.

Feeding Feed occasionally with a cactus and succulent fertiliser during the growing season.

Other maintenance tasks Wipe the leaves regularly to remove dust.

Danger signs Plants experiencing root rot will develop

yellow or brown soft patches on the leaves.

Toxicity Mildly toxic to humans and pets.

Display Some Sansevieria growers – notably Angel Ramos from Hawaii – have even had success using snake plants as bonsai material, although this may be more successful with the smaller, bird's nest types. Snake plants look good massed in a trough, planted individually in terracotta or glazed china, or pretty much any container provided there is some drainage. Make sure the pot is heavy so the plant does not topple over.

Cultivars This species of Sansevieria seems prone to spontaneous mutations that provide interesting varieties for the houseplant trade: there are dozens of very different looking varieties to choose from, such as the silvery, elegantly slender 'Bantel's Sensation', 'Twisted Sister' with its curious corkscrew leaves, the squat bird's nest rosettes of 'Hahnii' and the variegated 'Golden Hahnii'. Many have been further refined by breeders to create stunning cultivars.

Also try . . . There are dozens of other Sansevieria species to try: if you wish to enjoy scented flowers, *S. francisii* and *S. 'Fernwood'* are good plants to try. *Sansevieria masoniana*, the whale fin, has dramatic, broad leaves, while *S. aubrytiana* 'Metallica' is popular for its silvery variegation.

Saxifraga stolonifera

STRAWBERRY SAXIFRAGE

Saxifragaceae

As students of houseplants, we face double woes when attempting to understand the pantheon of plant species grown indoors: if the seemingly unpronounceable scientific names don't trip us up, we may fall foul of the many common names that are just plain misleading. Botanists will explain that African violets, for instance, are not violets, lucky bamboo is not bamboo, and the ponytail palm is ... well you get the idea.

Saxifraga stolonifera is one such plant, dogged with many names that compare it to things it is not. My preferred common name is strawberry saxifrage, but you may also find it labelled strawberry geranium, strawberry begonia, roving sailor, mother of thousands, creeping rockfoil, old man's beard, flesh plant or beefsteak plant. Common names bubble up over time, and lean towards the whimsical, often simply referencing the more popular plant to which they bear a passing resemblance. Some of these names are shared with several other species, to add to the confusion. I imagine a nursery owner somewhere looked at *S. stolonifera* and thought, 'Well, that looks a bit like a begonia with its patterned leaves, except those runners remind me of a strawberry. So, we'll call it strawberry begonia.'

One of the most exquisite images I have come across of this plant was a picture by English artist Mary Delaney made in 1775, now held at the British Museum in London: a strawberry saxifrage is set against a black background, its leaves and flowers fashioned from a collage of coloured paper and watercolour paint as well as a dried leaf from the actual plant. It is inscribed with three Chinese characters and the translation 'old tyger's ear' (the Chinese name for the plant is tiger's (or cat's) ear grass or leaf). On the reverse, there is an inscription explaining that this translation was provided by 'Whanga at Tong'. This man – whose actual name was Huang Ya Dong – was one of the first Chinese people to visit Britain, travelling from Canton (now Guangzhou) while still in his twenties with the help of botanist John Bradby Blake. Relatively little is known about his life, but what we do know is that his knowledge spanned many subjects, from discussing china production with Josiah Wedgwood to corresponding about plants with British botanist Joseph Banks.

Huang Ya Dong may well have recognised this plant, because it grows wild in his homeland. *Saxifraga stolonifera*'s range extends from southern China to Korea, Japan, Taiwan and Vietnam. In Japan, it is known as *yukinoshita*, which translates as 'under the snow', seemingly referring to the plant's frothy white flowers, which bloom from May to July. Although houseplant growers tend to display the strawberry saxifrage up high to better appreciate its looks, it is a ground-hugging plant when growing wild, occupying damp, shady places: rocky crevices, riverbanks and anywhere with moisture and protection from direct sun. When content, it will spread to create a mat-like mosaic of plants via means of the stolons that inspired its scientific name: wire-thin stems that emerge from the base of the rosette of the plant and head out in all directions in search of a pocket of soil. This stolon – or runner – terminates in a baby plant that, given the right conditions, will root and make a new plant that's genetically identical to the parent. If no such spot is found, the young plant will simply bide its time and stay attached to the umbilical cord of the parent.

Although horticultural texts often dismiss the flowers of the strawberry saxifrage as 'insignificant', I find them charming. They are asymmetrical, with two delicate, white and pink long petals below and three short petals above. These petals are held on slender pinkish-red hairy stems that often arise en masse, and resemble a cloud of butterflies; Christopher Lloyd, the late but fabled gardener of Great Dixter in Sussex, was right when he wrote in his book *The Adventurous Gardener* that they give the plant a 'puckish' look.

The strawberry saxifrage was brought to Europe in the early 1770s and was first described to western science in 1774 by the botanist William Curtis, best known for founding *Curtis's Botanical Magazine*: he called it *Saxifraga sarmentosa*, a name that persists in some houseplant books and articles to this day. This species is an outlier in the genus Saxifraga: the name means 'rock breaker', thought to refer to the plants' tendency to dwell in crevices. There are around 440 species recorded throughout the northern hemisphere, but *S. stolonifera* belongs to the section known as Irregulares, a group of around 20 or so species from the same region with a similar penchant for moist conditions, and a tendency to sport asymmetrical flowers.

Today, strawberry saxifrage is often referred to as an old-fashioned houseplant, but it was already being considered somewhat passé by 1842. A writer known only as 'a London nurseryman' reported in the *Floricultural Cabinet* that year, 'Time was when the thread-like stolones [*sic*] of *Saxifraga sarmentosa* ... might be seen pendent from almost every casement, and no mean appearance did it afford; but it falls to the lot of vegetables, as of humanity, that fashion reigns among their ranks; and, in consequence, the natural pendent development of many is banished to give way to the more constrained style of planting.'

And yet someone must have kept growing this plant. Perhaps the key to its persistence is its powers of abundant reproduction, meaning a stock of baby plants was always on hand for anyone who admired it. Another wonderful quality of the strawberry saxifrage is it belongs to a small subset of the horticultural world: species that are happy to be grown either indoors, or outdoors in a temperate climate where the thermometer sometimes falls a little below freezing in winter, as it does where I live, in southern England. (Other members of this club include two species also profiled in this book: *Aspidistra elatior* [see page 13] and *Hedera helix* [see page 91].) It is perhaps not surprising that the strawberry saxifrage has been recorded occasionally growing naturalised as a garden escapee in Britain since 1928, where the climate is not dissimilar to that of its homeland. Doubts about the plant's tenderness have often popped up, however. The *American Garden* noted in 1819 that *S. stolonifera* 'has been generally treated as a greenhouse plant', but confirms that, since it survived outside in the severe winter of 1804-5, it therefore could be considered 'perfectly hardy'. Christopher Lloyd agrees, noting that the plant 'laboured long under a tender reputation which seems totally unmerited'. The Royal Horticultural Society, which has awarded this plant its Award of Garden Merit, hedges its bets and classes the species as H2 in terms of hardiness, which means that it is 'tolerant of low temperatures, but not surviving being frozen', suggesting a minimum temperature of 1-5°C (34-41°F). In my garden, it has definitely survived to at least −5°C (23°F). Suffice it to say, S. stolonifera is a plant that relishes cool or unheated rooms and despises hot, stuffy centrally heated areas in winter.

The strawberry saxifrage has a long history of use in traditional medicine, treating several conditions ranging from diarrhoea and frostbite to eczema and abscesses. One of the more unusual treatments involves burning the dried leaves in a bucket: the resulting smoke is wafted over sore and swollen haemorrhoids. Scientists are beginning to investigate the active ingredients of this plant, and discovering that it does have antimicrobial and antioxidant powers as its traditional uses suggest. Extracts of the plant are also used in Japanese cosmetics, presumably for its powers as an antioxidant. For the moment, I am far more minded to test out the other practical application of this plant. In Japan, *yukinoshita* is classed as *sansai*, which means mountain vegetables: other notable species include butterbur (*Petasites japonicus*, known as *fuki*) and even the much maligned Japanese knotweed (*Reynoutria japonica*, known as *itadori*). Young tender leaves of *yukinoshita* are usually served as part of the dish tempura, where *sansai*, cultivated vegetables and seafood are lightly battered and deep fried.

CARE GUIDE

Light These are shade plants outside, but indoors, they will do well in a brightly lit room out of direct sunlight.

Temperature These make good plants for cooler or unheated rooms: porches, conservatories and bathrooms. They can survive an occasional freezing, but do better if kept to above 5°C (41°F), although the cultivar 'Tricolor' is less cold tolerant and therefore only suitable indoors. Frazzle them in a hot room with dry air and this will be enough to bring a swift end to your Saxifrage ownership.

Water They like a steady supply of moisture around their roots and will protest by wilting and crisping if too dry. Try a self-watering pot.

Humidity 50% humidity should be adequate.

Pests and diseases Any of the usual houseplant pests can strike – red spider mites, scale, thrips and mealybugs – but aphids seem to be the most prevalent pest, infesting new tender growth in spring in particular, and when temperatures are higher than the plant likes. The major pest of *S. stolonifera* when grown outside is the vine weevil, whose larvae make mincemeat of the roots. If you take plants outside in warmer weather, it is worth checking their rootballs before bringing them back inside. The culprit is a c-shaped, white larva around 1 cm long: remove and destroy any you find and, if you wish to be doubly sure, treat the plants with the biological control nematode *Steinernema kraussei*.

Substrate Regular houseplant potting mix or John Innes No. 2; if you tend to let plants dry out too much, add a handful of vermiculite to improve water retention.

Propagation This is one of the great joys of this plant. There are a couple of ways of dealing with the baby plants. You can loop back stolons carrying baby plants into the substrate of the parent plant or rest in a separate pot of potting mix. Either way, pin down the end of the stolon with an unfurled paper clip. They will root and, once established, the stolon can be snipped away. Strawberry saxifrage is what I'd call a short-lived perennial: even with the best of care, plants start to deteriorate after a few years, and it is wise to keep propagating so you have a stock of up-and-coming plants to replace the parent once it is past its best.

Feeding Fertilise regularly when in active growth with a foliage houseplant feed.

Other maintenance tasks Masses of stolons and their babies can start to look messy, especially if the stolons are broken, which causes them to regrow with branches. Snip wayward stolons off as close to the base as possible.

Danger signs Yellowing leaves usually indicate too much sun, or too much water around the roots.

Toxicity None known.

Display This is one plant that I would recommend putting into a plastic pot: terracotta tends to allow too much moisture to evaporate from the substrate. This plant really needs some height to allow its trailing stolons to drape down, giving the characteristic jellyfish look. A high shelf works, or a hanging planter of some kind. Some people recommend cutting off the flower stems as they emerge, particularly if they prefer to look at the foliage, but I leave them on – when flowering, you may wish to move the plant to eye level to enjoy the flowers.

Cultivars There has not been a huge amount of breeding work done on this species, but there are a few cultivars worth looking out for. My absolute favourite is 'Tricolor', with its leaves irregularly edged in cream and a lipstick pink tinge to all parts of the plant, giving way to bright pink leaf backs. It is more tender than the species, and a little trickier and prone to suddenly dying off. There are several other cultivars worth trying, particularly the large-leaved 'Nezu Jinja' which was collected from the Nezu district of Japan by modern-day plant hunter Cedric Basset.

Also try . . . Once you have collected all the cultivars of this plant, branch out into another species from the Saxifragaceae, this time indigenous to the west coast of North America. *Tolmiea menziesii* has its own curious way of reproducing, by growing baby plantlets at the junction of the leaf stalk and the leaf itself.

Spathiphyllum

PEACE LILY

Araceae

A confession: this plant almost did not make the cut for this book. I was genuinely concerned the best anecdote I would be able to offer was the time my mother mistook a spray bottle of weed-killer for leaf shine and ended up strafing her precious peace lily with poison. (Sorry, Mum.)

It was a piece of 1970s synthesiser music that finally convinced me the peace lily deserved a chapter. 'Swingin' Spathiphyllums' is the seventh track on *Mother Earth's Plantasia*, an album of early synth music released in 1976. The back cover features a cartoon of a gargantuan peace lily, so large there is a man sitting on the pot rim, smiling with his hands clasped and feet dangling in midair. Other tracks have similarly planty titles, including 'Music to Soothe the Savage Snake Plant' and 'You Don't Have to Walk a Begonia'. The 1970s was the last decade to experience a houseplant boom: the Mother Earth in the album's title was the Los Angeles plant shop that set the mould for the houseplant boutiques springing up across the world today.

Owners Joel and Lynn Rapp opened their store on Melrose Avenue in West Hollywood in 1970, after Joel's career writing for TV sitcoms faltered. The Rapps' slick marketing and showbusiness connections saw imitators popping up around southern California: a planty record was the perfect way to promote their store. The Rapps commissioned music for *Plantasia* from composer Mort Garson, who used a Moog synthesiser – an early entrant to the electronic music scene – to create music for you *and* your house-plants. According to the sleeve notes, 'every pitch is scientifically designed to affect the stomata, or breathing cells of your plants, opening them ever so slightly wider and allowing them to breathe ever so slightly freer and thus, grow ever so slightly better.' This was complete bunkum, but a brilliant strategy for promoting the whole Mother Earth brand: you could only buy the record in Mother Earth, via mail order through the Rapps' syndicated newspaper column or when buying a Simmons mattress from department store Sears (possibly the oddest marketing freebie of all time). The record became a cult classic and was reissued in 2019, which is when I first came across 'Swingin' Spathiphyllums' and became intrigued anew about this bog standard houseplant. Peace lilies tend to be the plant people send when a loved one dies; they also act as leafy filler in mall planter troughs and decorate desolate hotel reception desks across the globe. Even the surge of interest in the aroid family that began around 2018 did not offer much of a boost in popularity.

The name peace lily seems to stem from their white flowers, which resemble those of the calla lily (Zantedeschia species), which is also a member of the Araceae rather than a true lily. Nevertheless, peace lilies seem to have co-opted the symbolism of actual members of the lily family, which have been associated with death since the ancient Greeks, making them the go-to funeral plant. Like most aroids, the part of the Spathiphyllum we generally dub the flower is made up of two parts: the spathe – a large modified leaf in the shape of a white hood; and the spadix – a fleshy spike crowded with tiny bumps that make up the true flowers. Spathiphyllum literally translates as 'leaf spathe'.

You may notice that the title of this chapter refers only to the genus Spathiphyllum, rather than a particular species: that is because the vast majority of the plants on sale today are hybrids and cultivars with DNA from a number of different species within the genus. Some of these were created from traditional breeding techniques, while others were developed via tissue culture techniques. There are around 50 species in the genus, and most hail from Central and South America, with a couple from South-east Asia. The species you are most likely to have heard of is *S. wallisii*, because this is the plant most often mentioned in houseplant books and the one most often sold as a species in the past, but the others include *S. cannifolium*, *S. floribundum*, *S. patinii* and *S. cochlearispathum*. Peace lilies usually grow in the ground in the shady undercanopies of lowland forests, and like 'wet feet', as horticulturists say. That means they grow on banks of rivers and streams, where their roots can quest into the water, or in swampy areas; sometimes they grow as lithophytes, with their roots anchored to rocks in damp places.

In the wild, Spathiphyllums flower sometime between May and November, depending on the species: plants in cultivation are always offered for sale when in flower, whatever the month. In the 1980s, plant researchers at the University of Florida discovered that spraying peace lily leaves with the plant growth hormone gibberellic acid induced flowering within 70 to 100 days. Gibberellic acid is widely used to manipulate harvests of citrus and other fruits, increasing their setting rate and yield. The introduction of gibberellic acid treatments for growing Spathiphyllum also explains why the number of hybrids and cultivars on the market has exploded in the last few decades, as breeders can induce flowering – and therefore conduct cross-pollination experiments between different plants – at any time. Once the gibberellic acid has worn off from a new plant, it will stop blooming for a while; the spathes tend to turn green before they die back. Provided conditions are right, your plant should rebloom after a few months.

If you ever get to see Spathiphyllums blooming in the wild, keep an eye out for bee visitors: these are most likely male euglossine bees, sometimes known as orchid bees for the other flowers they pollinate. Like many modern roses, much of the scent has been bred out of cultivated Spathiphyllums, but wild ones are worth a sniff: S. cannifolium's scent has been described as spicy soap or clove, whereas S. wallisii has a whiff of lemon. These scents are tailored to draw in the bees, who put the organic compounds they contain to a very particular purpose. Using specially adapted hairs on their legs, the bees scrape away at the essential oils on the spadix to collect and store these on glands on their legs (pollination biologist Stephen Buchmann describes these glands as 'almost like a cowboy's saddlebags'); in the process, they pick up pollen to transfer to other flowers. Male bees spend weeks gathering a specially tailored palette of scents from peace lilies, orchids and even fungi and sappy wounds on trees, then gather for displays known as 'leks' where they buzz loudly and release the chemicals they have carefully collected into the air to attract females to mate.

This is not the only creature that seeks out Spathiphyllums: if you are lucky enough to find S. wallisii growing in its home of Colombia, you may also catch sight of the amber phantom butterfly

(*Haetera piera*) laying its eggs on the leaves. This butterfly flutters low through deeply shaded vegetation rather than seeking the sun, and is usually spotted at dusk feeding on fruits rotting on the forest floor. Its transparent wings glow a soft orange, and two eyespots at the edges of the lower wings help camouflage it against predators. Its brown, horny-headed caterpillars feed on the peace lily's leaves then pupate and emerge around two weeks later as adults.

The various Spathiphyllum species used to breed modern peace lilies were first collected by western botanists in Central and South America from the mid- to late 1850s, and they were popular as 'stove plants' grown in tropical greenhouses. Although they were popular as florist's flowers for the bouquets of American brides and debutantes as early as the 1930s and 40s, they do not seem to have become widespread as houseplants for the masses until the 1950s and 60s, when central heating became more commonplace, making it far easier to raise peace lilies indoors. By 1957, British houseplant nursery Rochford's had a whole greenhouse devoted to them, although British–American garden writer Thalassa Cruso called them 'a relatively new introduction for home growers' in her 1969 book *Making Things Grow*. The first references I could find to the common name peace lily was in the early 1970s: before that they were known as white sails, spaths or simply Spathiphyllum.

The 1980s brought more good news for the peace lily. Dr Bill Wolverton's research for NASA on houseplants' potential to remove air pollutants listed *S*. 'Mauna Loa' as one of the most effective plants to remove benzene, formaldehyde and other pollutants. As I explain in more detail in the *Chlorophytum comosum* chapter (see page 37), adding a houseplant to a room is far from the most efficient way to clean the air, but that hasn't stopped cementing Spathiphyllum's reputation as a living air filter on thousands of webpages. There is one other notable Spathiphyllum trend to note. In the early 2000s, American newspapers reported a craze for so-called War and Peace arrangements. These consisted of a clear glass vessel filled with water above a bed of brightly coloured gravel, a peace lily sprouting from the top, while among the roots swam a betta fish (also known as the Siamese fighting fish). Fortunately this fad seems to have petered out as our understanding of the care needs of bettas has improved: a vase of water is a poor substitute for a heated tank with an air filter.

If you take the 'war' out of the setup, though, there is no reason why a peace lily cannot be grown in a vase of water, provided you add nutrients regularly. This is the surprising truth about peace lily cultivation, and perhaps the ultimate reason for their success: they like moisture around their roots, unlike so many other houseplants that will quickly keel over in the same conditions. Given that many growers new to houseplants drown their plants, it renders them a popular beginners' plant. Many people grow them successfully by submerging their roots into a freshwater fish tank; a self-watering pot also works well, and cuts down on the amount of time you need to spend watering.

There is one more potential peace lily story on the horizon. A plant microbial fuel cell is an emerging form of green energy: plants capture sunlight through photosynthesis, while microbes on their roots break down organic matter and turn it into bioelectricity. The research is in its infancy, but a paper published in 2021 by Korean scientists Kei Jung Kwon and Bong Ju Park seems to offer hope that peace lilies could double up as ornamental plants with a practical purpose. Peace plants as batteries? I cannot think of anything better.

CARE GUIDE

Light They will burn if placed with your cacti in a big south-facing window in summertime, but other than that, they will survive in most settings. If your plant is failing to bloom, try giving it more light.

Temperature Warmth is essential: 20°C (68°F) and above is ideal, but no harm will be done if your plant experiences short spells of temperatures down to 14°C (57°F). Avoid cold draughts.

Water Peace lilies can wilt dramatically when denied water: they will perk up quickly once given a drink, but the cumulative stress of repeated drought will start to show on a plant eventually, so avoid letting the soil dry out completely. Wilting can also be a sign of waterlogged roots, so check the rootball carefully before watering.

Humidity Peace lilies enjoy humidity of at least 50%.

Pests and diseases Red spider mites only tend to strike this plant when it is already stressed through poor conditions.

Substrate Spathiphyllums do well in self-watering pots and hydroponic systems. Either choose a moisture-retentive substrate and water less, or opt for a more free-draining, airy mix, and water generously. Most peace lilies are sold in the former, so if you wish to repot, try combining a ratio of two thirds houseplant potting mix with a third of fine orchid bark, perlite or leca.

Propagation Peace lilies grow from rhizomes, a swollen horizontal underground stem, and one plant can be turned into several by division. Select a section of rhizome with at least one bud – known as a growth point – on it and a few leaves. Cut it away with a clean, sharp knife and pot up separately, positioning the rhizome just under the substrate surface.

Feeding Continue feeding regularly with any general houseplant fertiliser when the plant is in active growth.

Other maintenance tasks Once the flowers have faded and turned brown, cut them away close to the base of the plant.

Danger signs Leaves will curl inwards at the sides when the plant is experiencing drought stress, sometimes accompanied by too much sunlight. A mass of leaves turning yellow all at once is usually due to waterlogging.

Toxicity The pollen causes an allergic reaction in some people. Toxic to pets and people.

Display I would suggest being as radical as possible with your display of peace lilies, to avoid that 'hotel reception desk' look. That might take the form of a brightly coloured pot, or combining with contrasting foliage of other aroids in a mixed container; one climbing, another trailing.

Cultivars Peace lilies come in a range of sizes, from the tiny 'Power Petite' to the enormous 'Sensation', which can reach 180cm tall. There are variegated peace lilies for those who must have their greens served with a side of cream, including white-splashed 'Diamond' and 'Silver Cupido', whose leaves are green veined between misty white, and 'Silver Streak', which has a white midrib to the leaves. 'Golden Delicious' has yellow-green leaves the colour of the apple.

Also try . . . *Aspidistra elatior* (see page 13) is similar looking, minus the white flowers, but with the advantage of more drought tolerance.

Zamioculcas zamiifolia

ZZ PLANT

Araceae

In the early 1800s, Mare Street, Hackney, a few miles north of the River Thames in London, was a quiet backwater far from the bustle of the city. And yet a steady stream of visitors came to wonder at the Grand Palm House, Hackney Botanic Nursery's tropical greenhouse. It was built in 1818, 24 years before Kew's celebrated Palm House, and was packed with tropical plants from all over the world, all kept warm by an innovative steam system and watered with a network of pipes that showered the specimens below. As well as more than 100 species of palms, the glasshouse displayed hundreds of orchids (the nursery's catalogue of 1838 listed more than a thousand species) and a banana tree that almost reached the roof. This wasn't just a pretty display: it was a warehouse, as well as a cathedral to botanical imperialism. The value of this living inventory was estimated at £200,000 in 1828 - around £16 million in today's money.

One of the plants that would have been found in the Grand Palm House around this time was *Zamioculcas zamiifolia*, then known as *Caladium zamiæfolium*. This may come as a surprise, because it is usually described as being a recent addition to the houseplant market. The species was first recorded in print in 1828, in volume 15 of *The Botanical Cabinet* compiled by George Loddiges and his son Conrad, the owners of the nursery in Hackney. *The Botanical Cabinet* was the early 1800s combination of today's Sarah Raven nursery catalogue and a 'plantfluencer' Instagram feed: an aspirational publication that showed off plants to be purchased at the Hackney establishment. The Loddiges (or rather Conrad, as George was dead by this time, even though his name was still on the title page of this volume) wrongly listed *Z. zamiifolia* as coming from

185

Brazil. I'd speculate this is because John Forbes, the 23-year-old plant hunter who is recorded as having collected this species from its actual home in East Africa, had visited Brazil the year before. Forbes did not make it back to England with his haul of specimens, instead dying of malaria during a trip along the Zambezi in Mozambique in August 1823.

To discover more about this plant, we must travel back even further, to around 42 million years ago. Scientists interrogating records of fossilised pollen grains found on the eastern side of Africa have established that this is where and when Zamioculcas first appeared, thus earning this genus the title of 'fossil plant'. And still today Zamioculcas grows wild in tropical East and Southeast Africa, including Kenya, Malawi, Mozambique, South Africa, Tanzania and Zimbabwe. It can adapt to life in a range of terrains, growing as an understorey plant in coastal forests situated on old sand dunes, on savannahs and in forests, often forming dense thickets. Nutrient-poor soil is no problem for this plant: it thrives in fast-draining soil and rocky ground, perhaps explaining why in Swahili, one of the languages spoken where this plant grows, it is known as *mwendanavilima*, meaning 'goes with the hills'. This territory is also home to other drought-tolerant species that have been popularised as houseplants, including various species of snake plants (Sansevieria) and *Cissus rotundifolia* (grape leaved ivy).

The ZZ is a member of the aroid family, but it boasts a number of characteristics that mark it out as a bit of an aroid oddball. For a start, it is monotypic, meaning there is just one species – *Z. zamiifolia* – within the genus. It is the only aroid that performs CAM, a special form of photosynthesis that allows the plant to cope with hot, dry conditions (see page 47 for a full explanation). It is the only aroid that botanists agree can be designated as a succulent plant. And it is one of very few aroids to grow pinnate leaves: there is no main stem; rather the fleshy petioles (leaf stalks) emerge directly from a fleshy underground stem called a rhizome, which can reach 1m long. Along the petioles are a series of paired leaflets, arranged in a feather-like formation. The ZZ plant is marketed as a foliage houseplant and rarely flowers indoors, but it does bloom in nature, producing the corncob-shaped flower spikes that are a feature of most aroids. Admittedly these are relatively small and pop up close to the base of the plant, offering up less drama than the likes of *Monstera deliciosa* (see page 123), and no delicious edible fruit, either.

During seasonal droughts, the ZZ has a couple of coping mechanisms: it draws on the water and nutrients stored in its rhizomes, but in extreme droughts it will jettison its leaflets, leaving behind just the succulent petioles to reduce water loss as it waits for conditions to improve. These leaflets do not necessarily go to waste, however. The ZZ plant is the only member of the aroid family that can propagate itself from leaflets. When they fall, they may stay alive long enough for rains to arrive and prompt the growth of a swelling at their base within a few weeks – the start of a new rhizome and thus a new plant.

The appearance of *Z. zamiifolia* in a London glasshouse and in the Loddiges' publicity material did not herald the start of its rise to fame in the world of houseplants: that took almost two centuries. The ZZ sank back into obscurity, until botanist Heinrich Wilhelm Schott moved it to the new genus Zamioculcas in 1856. It popped up again in 1872 in *Curtis's Botanical Magazine*, as 'this most curious genus' *Z. loddigesii*. This periodical printed a botanical illustration of the species, reporting that specimens had been brought to Kew from Zanzibar by John Kirk in 1869. (Zanzibar covered a much larger area then: it encompassed the island of Zanzibar, and the coastal regions of Tanzania and Kenya.) Kirk spent two decades as the British vice consul in Zanzibar from 1866, during which time he sent many specimens back to Kew Gardens, as well as campaigning for the end of the slave trade in East Africa. Prior to that, he accompanied Dr David Livingstone as a botanist on his second Zambezi expedition. Thankfully for Kirk, he did not set out from Zanzibar with Dr Livingstone on his final, fatal expedition in search of the source of the Nile. In 1905, German botanist Adolf Engler gave the plant its current scientific name of *Z. zamiifolia*, referencing its resemblance to another group of ancient plants, the cycad genus Zamia, best known to houseplant growers via the species *Zamia furfuracea*, the cardboard palm.

It was the mid-1990s when the ZZ plant began to become a popular houseplant. In the intervening 120 or so years since Kirk's specimens arrived at Kew, the species is referenced in a few specialist books on tropical plants, most notably in German–American botanist Alfred Byrd Graf's *Exotica: Pictorial Cyclopedia of Indoor*

Plants, first published in 1958. In 1986, *The Indoor Garden*, a periodical from the (now seemingly defunct) Indoor Gardening Society of America, called the ZZ a 'crazy aroid with long stalks and paired leathery leaves which root like African violet leaves'. In 1996 Dutch nurseries started to propagate the ZZ in earnest; I have yet to track down the individual who decided to test out its potential as a mass market houseplant, although rumour goes that this mystery person found a plant in a ditch in Madagascar: this seems unlikely given this island is not part of the species' range. Nevertheless, it quickly took hold as a supremely popular specimen, prized for its toughness and its architectural looks: in 2002, the Florida Nursery, Growers and Landscape Association declared the species Plant of the Year. The ZZ has been embraced around the globe in the last three decades, each culture attaching a different symbolism to it. A study of *Z. zamiifolia* growth in Cuba reported that there it is known as the plant of Che Guevara, because it grows abundantly in the gardens around the monument to the legendary Marxist revolutionary. In China, it is considered an auspicious plant and called *Jin qian shu* in Mandarin – which means golden money tree – because its rounded deep green succulent leaves remind people of the jade plant (see page 45). Across all cultures, though, its main quality is its toughness. It's often described as unkillable or indestructible, hence one of its common names, the eternity plant. This makes sense, given its tenacity in the wild.

There's one other ZZ legend that needs tempering. Houseplant books inevitably refer to this species as being toxic to humans, and there are various claims dating back to the 1990s that just touching this plant will give you cancer. There is no scientific evidence to show the cancer claims are true, and scant evidence to show the plant is particularly toxic. Indeed, the studies that have been carried out seem to indicate that the ZZ plant's toxicity levels are low or non-existent. That's not to say you, your children or your pets should turn a Zamioculcas into a tasty snack: even though ingesting a plant won't kill you, it may still taste revolting, cause severe irritation and land you in need of medical attention.

Recent research indicates that the ZZ may be more than just an aroid anomaly. Two examples: extracts from ZZ stems have been found to be effective against bacterial infectious diseases. And when it was tested alongside fellow houseplants, the corn plant (*Dracaena fragrans* 'Golden Coast') and the peace lily (*Spathiphyllum* 'Verdi'), it was found to be equally good at removing the air pollutant nitrogen dioxide from the atmosphere in a poorly ventilated indoor space, even if you only have five plants.

CARE GUIDE

Light The ZZ can cope with relatively deep shade, but it will subsist rather than thrive in the darkest corners of a north-facing room. If it is too dark to read a book easily, it's way too dark for your plant. It will not burn in morning sun, but avoid a south-facing windowsill at the height of summer.

Temperature Do not allow the temperature to fall below 12°C (54°F).

Water Although the ZZ plant can cope with periods of drought, it will grow better if watered regularly: the higher the light levels, the more it will drink. Never allow it to sit in water, and ease off watering when days become shorter in winter.

Humidity Dry air is no barrier to success with the ZZ plant.

Pests and diseases Delightfully few pests bother the ZZ; but it could be struck by any of the usual range of suspects, particularly mealybugs and scale insects.

Substrate Try to emulate the sandy or rocky conditions of its native habitat by adding extra drainage material to houseplant potting mix or John Innes No. 2 at a ratio of about 1:3; leca or horticultural sand is ideal.

Propagation Division works well for larger specimens: remove the plant from the pot and tease apart the rhizomes, potting them up separately. If you want more plants but are prepared to wait longer, try rooting individual leaflets. Gently pull them away from the petiole at the base, and place the cut end in gritty potting mix. Keep covered with glass or a clear plastic bag somewhere warm and light: the leaflets will produce their own tiny rhizomes within a few weeks. Even sections of leaves can be propagated, but this may make the task too time consuming for most.

Feeding A regular application of half strength houseplant foliage feed will be sufficient.

Other maintenance tasks
Regular wiping of the leaves – or

showering down the plant with soft or distilled water – will stop them from getting dusty.

Danger signs If leaflets start to fall, check the substrate: long periods of drought will cause the plant to go dormant, but waterlogged roots can also cause foliage loss.

Toxicity Doubtful toxicity to humans, but certainly not edible. Usually classed as toxic to pets.

Display ZZ plants look impressive massed in a trough. The rhizomes can grow so vigorously that their roots can 'burst' out of pots, so plastic may be preferable.

Cultivars A few cultivars of the ZZ plant have come to market, including the compact 'Zamicro' and 'Zenzi'. Probably the best-known cultivar is 'Raven', whose leaflets start out emerald green but age to deepest purple. Variegation fans may wish to seek out 'Lucky White' with yellowish tips to the stems, but note that, like most variegated plants, this is slower growing than its plain green counterpart.

Also try . . . *Gonatopus boivinii* or giraffe's knees is a close relative of the ZZ plant: it used to be placed in the genus Zamioculcas, comes from the same region of tropical East Africa and is another unfussy plant for indoor culture, although it is a lot less widely available. It's named after the bulge on its stem that looks rather like that on a giraffe's leg.

Further Reading

Abbott, Daisy T., *The Indoor Gardener*, University of Minnesota Press, 1939

Adams, W. H. Davenport, *Wonders of the Vegetable World*, T. Nelson and Sons, 1867

Allan, Mea, *Tom's Weeds: Story of the Rochfords and Their House Plants*, Faber & Faber, 1970

Altman, Boy, *Indoor Ferns: Caring for Ferns*, Rebo Productions, 1998

Bown, Deni, *Aroids: Plants of the Arum Family*, Timber Press, 2000

– *Aroids: Plants of the Arum Family* (second edition), Timber Press, 2010

Buchmann, Stephen, *The Reason for Flowers: Their History, Culture, Biology, and How They Change Our Lives*, Scribner, 2015

Burchell, William J., *Travels in the Interior of Southern Africa, Volume 1*, Longman, Hurst, Rees, Orme and Brown, 1822

Candeias, Matt, *In Defense of Plants*, Mango Media, 2021

Carter, S., Lavranos, J. J., Newton, L. E. and Walker, C. C., *Aloes: The Definitive Guide*, Kew Publishing, 2011

Chahinian, B. Juan, *The Splendid Sansevieria*, Trans Terra Publishing, 1986

Condit, Ira J., *Ficus: The Exotic Species*, University of California, Division of Agricultural Sciences, 1969

Crow, Juliana, *Your Indoor Plants: From Aspidistra to Zalacca*, Weidenfeld & Nicolson, 1952

Cruso, Thalassa, *Making Things Grow*, Alfred A. Knopf, 1969

Culpeper, Nicholas, *Culpeper's Complete Herbal: Illustrated and Annotated Edition*, Arcturus Publishing, 2019

Cunningham, Scott, *Cunningham's Encyclopedia of Magical Herbs*, Llewellyn Publications, 2012

D'Amato, Peter (revisited by), *The Savage Garden*, Ten Speed Press, 2013

Diaz, Juliet, *Plant Witchery*, Hay House, 2020

Earle, Roy A. and Round, Janice E., *Lithops in Habitat and Cultivation*, Lavenham Press, 2021

Garrett, J. T., *The Cherokee Herbal: Native Plant Medicine from the Four Directions*, Bear & Company, 2003

Graf, Alfred Byrd, *Exotica: Pictorial Cyclopedia of Indoor Plants*, Roehrs Company, 1958

Griffin-King, June, *Indoor Gardening* (Ladybird Series 633), Ladybird Books, 1976

Griffith Jr, Lynn P., *Tropical Foliage Plants: A Grower's Guide*, Ball Publishing, 2006

Heine, Bernd and Legère, Karsten, *Swahili Plants: An Ethnobotanical Survey*, Rüdiger Köppe, 1995

Hessayon, Dr D. G., *Gold Plated Houseplant Expert*, Ebury Press, 1989
— *The House Plant Expert*, PBI Publications, 1980

Horwood, Catherine, *Potted History: The Story of Plants in the Home*, Frances Lincoln, 2007

Hoshizaki, Barbara Joe, *Fern Grower's Manual*, Alfred A. Knopf, 1975

Jenkins, Dorothy H. and Van Pelt Wilson, Helen, *Enjoy Your House Plants*, M. Barrows & Company, 1948

Kramer, Jack, *Begonias as Houseplants*, Van Nostrand Reinhold Company, 1976
— *Ferns and Palms for Interior Decoration*, Grendel Books, 1972
— *Indoor Trees*, Val Waring, 1978

Lawrence, Gale, *A Naturalist Indoors: Observing the World of Nature Inside Your Home*, Phalarope Books, 2000

Lee, David, *Nature's Fabric: Leaves in Science and Culture*, University of Chicago Press, 2017

Lloyd, Christopher, *The Adventurous Gardener*, Penguin Books, 1985

Loewer, Peter, *The Evening Garden*, Hungry Minds Inc., 1993

Loudon, J. C., *The Green-House Companion*, Whittaker, Treacher and Co., 1832

Mabey, Richard, *The Cabaret of Plants: Botany and the Imagination*, Profile Books, 2015

Martin, Tovah, *Once Upon a Windowsill*, Timber Press, 1988

Mayo, S. J., *Flora of Tropical East Africa - Araceae*, Routledge, 1985

Mwachala, Geoffrey and Mbugua, Paul Kamau, *Flora of Tropical East Africa: Dracaenaceae*, Royal Botanic Gardens, Kew, 2007

Nelson, E. Charles and McKinley, Donald, *Venus's Flytrap* (Natural History Series), Boethius Press, 1991

Nelson, Gil, *The Ferns of Florida: A Reference and Field Guide*, Pineapple Press, 2000

Ollerton, Jeff, *Pollinators & Pollination*, Pelagic Publishing, 2021

Parmenter, Ross, *The Plant in My Window*, TY Crowell Company, 1949

Perry, Frances, *Beautiful Leaved Plants*, Scholar Press, 1978

Phillips, Henry, *Floral Emblems*, Saunders and Otley, 1825

Reynolds, Gilbert Westacott, *The Aloes of Tropical Africa and Madagascar*, Swaziland: Aloes Book Fund, 1966

Rochford, Thomas and Gorer, Richard, *The Rochford Book of Houseplants*, Faber & Faber, 1961

Rowley, Gordon, *Succulent Compositae: A Grower's Guide to the Succulent Species of Senecio and Othonna*, Strawberry Press, 1994

Sayers, Edward, *The American Flower Garden Companion*, J. A. James, 1846

Schiebinger, Linda and Swan, Claudia, *Colonial Botany: Science, Commerce, and Politics in the Early Modern World*, University of Pennsylvania Press, 2007

Simons, Paul, *Potted Histories: How to Make House Plants Feel at Home*, BBC Publications, 1996

Solman, David, *Loddiges of Hackney: The Largest Hothouse in the World*, The Hackney Society, 1995

Sparke, Penny, *Nature Inside: Plants and Flowers in the Modern Interior*, Yale University Press, 2021

Steele, Joelle, *Interior Landscaping Dictionary*, John Wiley, 1992

Stover, Hermine, *The Sansevieria Book*, Endangered Species Press, 1983

Sturgeon, Andy, *Potted*, Conran Octopus, 2001

Success With House Plants, Reader's Digest, 1979

Tan, Hugh T. W. and Xingli, Giam, *Plant Magic: Auspicious and Inauspicious Plants from Around the World*, Marshall Cavendish, 2008

Teltscher, Kate, *Palace of Palms: Tropical Dreams and the Making of Kew*, Picador, 2020

Titchmarsh, Alan, *The Hamlyn Guide to Houseplants*, Littlehampton Book Services, 1982

Torre, Dan, *Cactus*, Reaktion Books, 2017

— *Carnivorous Plants*, Reaktion Books, 2019

Utteridge, Timothy, *Kew Tropical Plant Identification Handbook*, Kew Publishing, 2015

van Wyk, Braam, *Field Guide to the Trees of Southern Africa*, Struik Nature, 2013

Vickery, Roy, *Vickery's Folk Flora: An A–Z of the Folklore and Uses of British and Irish Plants*, Weidenfeld & Nicolson, 2019

Wennström, Anders and Stenman, Katarina, *The Genus Hoya: Species and Cultivation*, Original, 2008

Whittingham, Sarah, *Fern Fever: The Story of Pteridomania*, Frances Lincoln, 2012

Wiersema, John Harry, *World Economic Plants: A Standard Reference* (second edition), CRC Press, 2013

Willis, Abigail, *The Remarkable Case of Dr Ward and Other Amazing Garden Innovations*, Laurence King Publishing, 2018

For a list of the key scientific papers used in researching this book, visit **janeperrone.com/legendsresearch**

Acknowledgements

If making a podcast about houseplants is a series of 100-metre sprints, writing a book about them is an ultramarathon. Fortunately I had a wonderful crowd of people to cheer me on and guide my way to the finish line. Huge thanks to all of the following people: may mealybugs forever forsake you.

To Rick, my biggest fan and patron, for the endless cups of tea while putting up with my impossibly messy desk and constant encroachment from my unruly plants. To Ellen and Fred, who knew when I needed a hug. And Wolfie the lurcher, who saved our mental health during the pandemic.

To my delightful and diligent collaborator Helen Entwisle, whose illustrations bring this book to life.

To all the people who read draft chapters and offered up their wise thoughts and comments, including Joe Bagley, Tom Bennet, Robbie Blackhall-Miles, Peter Blake, Memo C, Dr Matt Candeias, Doug Chamberlain, Ben Dark, Dr Leon van Eck, Stephen Ehlers, Sarah Gerrard-Jones, Leslie Halleck, Bobby Ho, Matthew Jackson, Mark Lashmar, Tony Le-Britton, Mercy Morris, Dr Jeff Ollerton, Philip Oostenbrink, Joan and John Perrone, Hannah Powell, Kash Prashad, Lauren Smith, Dr Paul Twigg, Dr Colin Walker and Dr Scott Zona.

To the many people who helped me track down information, locate obscure books and translate foreign texts, including Claude Barrère, Cédric Basset, Dr Peter Boyce, Dr Clint Carroll, Marco Cedeño Fonseca, Katherine Cook, Fiona Gilsenan, Ulrich Haage, Catherine Horwood, Mick Mittermeier, Dr Daniel Whistler, Dr Danielle Sands and Sally Savelle.

To Joelle Owusu-Sekyere, who first said yes to *Legends of the Leaf*, and all at Unbound who helped to make the book a reality, including DeAndra Lupu, Cassie Waters, Anna Simpson and Katy Guest.

To my assistant Kelly Westlake for being the most cheerful and resourceful right-hand woman, and the staff at the RHS Lindley Library who dug out obscure houseplant books for me to peruse and made it possible for me to visit even during the pandemic.

And finally, to all the listeners of *On The Ledge*: enjoy this book, because you made it happen. And if you find yourself hearing my voice in your head as you read, I can only apologise.

A Note on the Author

Jane Perrone is a horticultural expert, journalist and the host of *On The Ledge*, a podcast dedicated to houseplants and indoor gardening. She is a regular contributor to the *Guardian*, the *Financial Times* and *Gardens Illustrated*. She lives in Bedfordshire with her husband, two children, a dog called Wolfie and a home full of plants.

A Note on the Illustrator

Helen Entwisle is an illustrator and printmaker based in Cumbria, often found mixing just the right shade of ink or printing with her favourite squeegee. She has worked on many varied projects during her fifteen-year career, drawing and screen-printing for clients and for her Memo Illustration product line. As a big plant-lover herself, creating these houseplant illustrations to complement Jane's words has been a leafy dream come true.

Index

Unbound is the world's first crowdfunding publisher, established in 2011.

We believe that wonderful things can happen when you clear a path for people who share a passion. That's why we've built a platform that brings together readers and authors to crowdfund books they believe in – and give fresh ideas that don't fit the traditional mould the chance they deserve.

This book is in your hands because readers made it possible. Everyone who pledged their support is listed below. Join them by visiting unbound.com and supporting a book today.

With special thanks to Richard Holliman and Francesca Murray Rowlins for their generous support of this book

Francesca Ablett, Ayo Adepoju, Natalie Akenzua-Sanderson, Gabrielle Akers, Sally Akers, M Albespy, Denise Aldred, Marc Alexander, Kat Allender, Summer Alp, Evelin Andrespok, Jane Angell, Curt Anstey, Steven Appleman, Alexander Arbelaez, Rachel Arday, Liz Armitage, Kathy Armstrong, Catherine Arnold, Chris Arrowsmith, Jacquelyn Arsenuk, Sarah Ashlee, Adrian Ashton, Corinne Atherton, Andrew Atkinson, James Aylett, Natalie Bagnall, Victoria Bailey, Mary Baillie, Leah Baird, Hannah Baker, Jo Baker, Crystal Ball, Claire Bandfield, Christopher Bangs, Maria Baranowska, Alexander Barbour, Ross Barbour, Henriette Barchager, Julia Barnes, Sara Barnes, Emma Barton, Alison Bateman, Natalie Baum, Helen Baxter, Lottie Baxter, Emma Bayliss, Angela Bazan, Sarah Bell, Trent Bell, Mylene Belleau, Erin Bendle, Kathleen Bennallack, Katlynn Bennett, Simon Benson, Julie Beynon, Nikki Bi, Barrie Birch, Kimberly Black, MJ Black, Peter Blackburn, Julianna Blagg, Lorna Blair, Eve Bloom, Maureen Bocka, April Bogard, Twanna Bolling, Andrew Bolton, Jean Bond, Stacey R Booth, Kelli Bordner, Petra Bosnyák, Tess Botsis, A Boyle, Clare Boyle, Lydia Bradley, Mary Brady-Maguire, Juniper Branscomb, Chase Brestle, Paul Brewer, Anna Brisley, Robert Broad, Catherine Bromley, Charlotte Brooke, Douglas Broome, David Brown, Katie Brown, Natasha Brown, Brian Browne, Birthe Brüggemann, Suzanna Bruin, Celso Bruzadin, Kezia Bryant, Gregory Buchholz, Colin Buck, Rachel Buckley, Charles Budd, Kate Burrows, Eliza Burt, Claudia Buszta, Elaine Butler, Jonathan Bykowski, Stephen Byrne, Susan Cadzow, Matthew Cairns, Alexandra Campbell, Claire Capdevila-Wright, Stephanie Carroll, Zoë Cartlidge,

Emily Rose Cassidy, Joanna Castaneda, Cherie Catchpole, Ophélie Cattin, Kristina Causer, Bitzy Cave, Victoria Cawthorne, Anne Chacon, Paul Chandanabhumma, Baylor Chapman, Philippa Charatan, Lynn Charles, Emily Charlton, Theo Charnley, Darryl Cheng @houseplantjournal, Simon Cherry, Esther Choi, Anna Chojnacka, Amanda Christensen, Darcey Christian, Agamemnon Christodoulou, Noelia Chung, Mikal Ciccolini, Helen Clancy, Theresa Clancy, Anna Clarke, Ken Clarke, Gaby Clayton, Sara Clemmer, Lauren Clinnick, Nikki Coates, Tessa Cobley, Geoffrey Cochran, Natalie Cole, Rhiannon Cole, Cassidy Coleman, Trevor Collins, James Common, Nicki Conlon, Christine Cook, James and Jael Cook, Julia Cooke, Lin Cooke, Marissa Coolidge, Clifford Alfred Coppin, Michael Cordy, Tabitha Pryor Corradi, Janice Corriden, Rowena Cortez, James Cosgriff, Kaitlyn Cottrell, Andy Cowley, Emily Cox, Colleen Coyle-Levy, Lucia Craig, Mags Craig, Thomas Cranham, Kathryn Crews, Tessa Crocker, Jennifer Croskrey, Alison Cross, Karen Crowley, Julia Croyden, Kathrine Cuccuru, Kathy Cuckoo, Alastair Culham, Kyleleen Cullen Bartnick, Vickie Curtis, Karolina Czaplinska, Jenn D'Antonio, Maarten Daalder, Joanna Daborn, Charlotte Dadswell, Royal Daniel, Greg Danner, Jennifer Darby LeGrange, Lisa Darnell, Elizabeth Darracott, Debbie Davies, Gareth Davies, Lynwen Davies, Jane Davis, Tim Davis, Steven Day, Anabelle de Chazal, Marcela De la Pena, Lianne de Mello, Jennifer De Soto, Jon Dean, Erik DeBill, Sara Dee, Maren Deepwell, Rachael Dehe, Lisa Del Papa, Mandy Delmaine, Sarah DeLoach, Mike Diamond, Megan Dietz, Renee Dillon, Lisa Ding, Angela Doherty, Omar Domingo, Kevin Donnellon, Kellie Donovan, Sam Doolan, Lauren Dorosh, Katrin Doughty, Andrew Drake, Alison Drever, Robyn Drinkwater, Eleanor Driscoll, William J Drury, Steve Duckworth, Keith Dudleston, Julia Duffy, Lisa Dunne, Andrew Dunngalvin, Sophie Durlacher, Rob Dwiar, Scott Dwyer, Elizabeth Dyer, Siew Dyer, Jane Eastgate, Timea Edvi, Katherine Edward, Claire Edwards, Ross Edwards, Zosia Edwards, Stephen Ehlers, Edward Ellebracht, Joff Elphick, Imogen Ely, Sarah Ely-McCollum, Susanne Emde, Mia Enbom, Jennifer Endsley, Karoline Engl, Eileen Entwisle, Robin Ernst, Maria Failla, Tilly Fairlie, Ying Har Fan, Kathryn Fargher, Kayla Farrar, Marchelle Farrell, Aron Feher, Christine Felstad, Angela Ferguson, Leonora Ferguson, Darren Ferris, Helen Fielder, Lynn Fiorentino, Jane Fitts, Marisa Flanagan, Ellen Flatley, Lisa Fletcher, Clare Foggett, Katherine Folkard, Jessica Forbes, Colin Forrest-Charde, Carole Forsythe, Lucy Foskett, Alys Fowler, Clare Fowler, Heidi Fox, Janet Fox, Roz Fox-Bentley, Jo Frankham, Sarah Freeman, Marian French, Heather Frizzell, Cary Fulford, Hazel Fulton, Brittani Gabowitz, M Galbraith, Hilary Gallegos, Susie Galley, Mark Gamble, Ann Garascia, Eric Garcia, Jessica Gardner, Pamela Garnett, Macauly Gatenby, Vincent Gauci, Sarah Geromini, Sarah Gerrish, Jeannette Geyer, Camilla Ghazala-Saunders, Susan Gibert, Amy Gilbert, Julie Giles, Anna Glendening, Dena Gonsalves, Stella Good, Chantal Gordon, Emily J Gordon, True Gordon, Mina Gourlay, Katherine Grace, Rachael Graham, Karen Greaves, Juliette Green, Cathryn Gregor, Ross Gregory, Maria

Groen, Katy Guest, Katherine Gumbs, Chris Gunnell, Gozdem Gurbuzatik, Marta Gutierrez, Claudia Gutke, Selina Gynn, Stephanie Haddock, Abigail Hagen, Antony Hagues, Irene Hailes, Eva Haisova, Kiki Halaska, Sarah Hale, Andy Hall, Bea Hall, Jessica Hallett, Caroline Hallstrom, Summer Hamel, Maria Hammarlund, Kamille Hammerstrom, Sileny Han, Jade Hannah, Dave Hanson, Kyra Hanson, Liz Hanson, Mary Hardgrave, Karen Harding, Fay Hare, Geoff Hargreaves, Lee Harkness, Ashley Harper, Charlotte Harris, Jim Harris, Charles Hart, Anna Hartman, James Harvey, Lucy Haskell, Kjetil Haslum, Sall Hathaway, Amanda Haugen, Sandra Haurant, Meg Haver, Melinda Hawthorne, Nathan Heath, Lindsay Heller, Joe Hendershot, Suzanne Hendrix-Case, Elizabeth Heness, Glynis Hensley, Juan Manuel Herrera, Andrew Herrmann, Jeff Hess, Charlene Hewitt, Sasha Hewitt, Jo Hickman, Laura Higgins, Stephen Higgs, Jodie Hill, Kath Hindley, Natalie Hines, Bobby Ho, Donna Hodds, Helen Hodgson, Carola Holbeck, Adele Holgate, Donna Holland, Harry Edwards Holliman, Neville Hollingworth, Molly Hollman, Garth Holman, James Holmes, Carrie Hoppes, Davey Horne, Jane Horsman, Catherine Horwood Barwise, Susan Housley, Alice Howe, Amber Hrynczyszyn, Johnson Huang, Diego Huet, Amber Huff, Fiona Hughes, Anna Hummel, Caroline Humphreys, Janine Hunter Hall, Holly Hunter-Perrine, Sandra Hurford, Peter Huynh, Sylvia Huynh, Anita Iacobacci, @imperfectjungle, Rich Ingham, Alison Innes, Charlotte Iosson, Heather Irvine, Derek Irwin, Amanda Jackson, Chantal Jackson, Matthew Jackson, Richard Jackson, Zoë Jackson-Newbold, Amy Jacobs, Valerie Jacobson, Zoe James, Kiera Jamison, Stacia Janis, Nicola Jenkins, Jessica Jensen, Simon Jerrome, Jerry @jeryszgarden , Mariette Johansson, Zoe John, Jamie Lynn Johnson, Joshua Johnson, Caroline Johnston, Clive Jones, John Jones, Rebecca Jones, Sarah Jones, Eleanor Jones-McAuley, Karolina Kajak, Lukashwarie Karimzandi, Ada Karlsson, Martina Karlsson, Anna Karsten, Kathryn, Sonia Kay, Joanne Kaye, Ellen Kaye-Cheveldayoff, Sumiko Keay, Kathrin Kehl, Lee Kelman, Mim Kendrick, Devin Kennedy, Shaun Kenny, Lynda Kern, Jack Kerruish, Ruth Keys, Stefanie Kiell, Dan Kieran, Rosie Killick, Sam Kilpatrick, Kara Kimmel, Susan King, Brian Kirkwood, Kristina Klee, Daniel Kleinman, Tori Knollmeyer, Sarah Knowles, Roos Kocken, Michael Kowalski, Viktoria Kozlova, Martha Krempel, Helene Kreysa, Erin Kriss, Emily Kultgen, Patti Kunnmann, Sophie L, Misty La Peer, Mireia Lacort Lopez, Edith Laird, G Lally, Jason Lamountain, Shannon Lamoureux, Nicholas Land, Freya Langford, Caroline Langton, Hannah Lannon-Black, Elizabeth Larson, Mark Lashmar, Franca Lauria, Vanessa Lawrence, Abigail Lawson, Aaron Le Marquer, Tony Le-Britton, Mary Lee, Lia Leendertz, Calvin Lefebvre, Toni Leitgab, Paula Leslie, Karl Leung, Alison Levey, Kyle Lewis, Megan Lewis, Helen Li, Isabel Limb, Gillian Lingwood, Petal Linnie-Godden, Bec Lloyd, Eleanor Long, Nancy Longnecker, Marta Lopez Carrete, Annette Lovett, Lora Lowe, Melissa Lowrie, Felicity Lugton, Nicholas Lui, Cuong Luong, Kelly Lytton, Alison M-Hodgkiss, Rob MacAndrew, Paige MacKenzie, Skye MacKenzie, Alexis Mackie, Christopher MacPherson, Mandy MacRae, Savannah Madley,

Sharon Magner, Jessyca Malina, Ava Mallett, Laura Mandelson, Peter Manos, Jenny Maresh, Kacie Marie, Sandy Marks, Kate Marsden, Tonia Marsh, Julia Marshall, Jesús Martín Sánchez, María Victoria Martín Sánchez, Catherine Mason, Bryan Masters, Susanne Masters, Sarah May, Mairéad Mc Lean, Mary McCarthy, Molly McCarthy, Janice McCreary, Josh McCune, Liz McFall, Erica McGillivray, Megan Marie Mcginnis, Maureen McGoldrick, Jean McGrogan, Sarah McGuire, Natascha McIntyre Hall, John McKenzie, Hannah McLean, Kara McLellan, Melanie McLeod, Hayley Meaden, Joel Meador, Selam Mebrahtu, David Meeres, Alicia Merriam, Ed Middleton, Tamara Miljus, Rachel Miller, Ben Millett, Allison Mis, Bethan Miskell, Sarah Mitchell, John Mitchinson, An-Magritt Moen, Sonja Moffat, Alex Molton, Chris Moncrieff, Janet Montefiore, Lee Moody, Felipe Morfin Martínez, Kaitlin Morgan, Jane Moritz, Rosie Morland, Kristin Morrell, Cath Morris, Fionnuala Morris, Kirsty Morris, Barbara Morrison, Heather Morrison, Rebecca Morton, Helen Mosley, Joe Moss, Lucy and Julian Moss, Mark Moulang, Ellie Mowforth, L Moyle, Joerg Mueller-Kindt, Alice Munro, Jason Munshi-South, Peter Münster, Alyssa J Murphy, Angie Murphy, Susan Murphy, Marie Murray, Rhiannon Murray, Alice 'Monstera' Murrell, Vanda Naden, Sal Namboodiri, Caitlin Napleton, Caspian Nash, Jessica Naugle, Carlo Navato, Aaron Nazar, Lisa Needham, Deirdre Neff, Olivia Nelson, Paula Nelson, Jenny Newman, Matthew Newman, Lou Nicholls, Ian Nichols, Duncan Edward Nicol, Nikki the plant killer, Claire Noble, Lisa Nolan, Sandra Norvell, Sarah Norwood, Clara Notartomaso, Natasha Nuttall, Yann Ó hEireamhóin, Paul O'Hara, Mark O'Neill, Summer Rayne Oakes, Sharon Oldham, Janice Olivo, Genevieve Ollerearnshaw, Erica Olsen, Karen Olsoy, Elaine Orchard, Helen Orr, Caroline Ortego, Angela Osborne, Belinda Otas, Shara Ouston, Sarah Owen-Hughes, Joelle Owusu, Virginia Oxford-Fleming, Kevin Ozebek, Paul Palisin, Allyson Paradis, Alison Park, Julia Parker, Maria Parker, Courtney Parks-Sindergard, Erin Partridge, Ana Patino, Graham Pavey, Lorraine Percival, Rhonda Pereles, Kristina Perez, Ellen Perrone, Fred Perrone, Joan Perrone, John Perrone, Louise Perrone, Anthony Peterson, Josh Peterson, Thea Petrou, Beth Pfeifer, Michelle Phalp, Brittany Pharo Wacker, Beth Picton, Helen Picton, Anna Pietroni, Kirsten Platt, Carrol Plummer, Justin Pollard, Gina Porras Parral, Jacqui Porritt, Lynette Porter, Nathan Porter, Matt Potter, Mimi Prendergast, Yvonne Price, Vivienne Raftery, Jessica Rahardjo, Susan Rainey, Amanda Ramos Crockett, Emily Randall, Benjamin Ranyard, Yvonne Rassam, Iida Rauma, Ollie Ray, Rachel Rayner, Kristopher Read, Bethan Rees, Andrew Reese, Karen Reilly, Anna Reimondos, Marina Remington, Katie RG, Emma Rhind-Tutt, Charis Richardson, Marc Ridpath, German Rivera, Debbie Roberts, Sophie Roberts, Hilary Robinson, Katrina Roche, Stephanie Rockco, Laurie Rodriguez, Alma Rojas, Roz Rosewarne, Nancy Ross, Tomasz Roszkowski, Rachel Roszman, Jani Rounds, Sarah Rousseau, Soledad Royo, Annemarie Rundlett, Karen Ryan, Mary Ryan, June Saddington, Michelle Sadlowski, Raymond Salcedo, Joe & Kate Salmon, Samantha Salmon, Heidi Saltsman, Jennifer Sanquer-Mason,

Sarah The Plant Rescuer, Nis Sarup, Pauline Sato-Paulino, Lisa Scannell, Polly Schiffman, Naomi Schillinger, Cat Schroeter, Anne Schuetz, Barbara Segall, Amer Sehmi, Claire Selwood, Alaina Sepulveda, Jennie Serrano, Emily Seward, Bev Shalts, Teresa Shannon, Laurence Shapiro, Orlagh Sheils, Jeff Sheng, Larry Shepherd, Amy Shieh, Adam Shields, Alex Shields, Patricia Short, Johnny Sierra, Holly Simpson, Emily Singer-Ripley, Paula Siqueira, Julia Sittmann, Theodor Sjölin, Janetta Skarp, Tim Skellett, Caitlin Skinner, Jessie Slater, Naomi Slater, Irmani Smallwood, Miles Smaxwell, Rae Smedley, Ane Smith, Cassandra Smith, Charlotte Smith, Hannah Smith, Joanne Smith, Karen Smith, Lauren Smith, Xander Smith, Zita Soregi, Solange Souris, Teagen Southey-Faoagali-Latu, Ollie Spackman & Nele Piepenbrink, Neely Spicer, Grace Spitzer-Wong, Kayla Sprague, Rochelle St Jean, Jennifer Stackhouse, Amanda Stacpoole, Catriona Staddon, Morgan Staden, Steve Stafsholt, Lynn Stannard, Emily Starling, Nikolas Stasulli, Anne-Marie Stavert, Abby Steinbrook, Maria Steiner, Lisa Eldred Steinkopf, Craig Stephen, Carolyn Stephenson, Alison Stewart, Jane Stewart, Karen Stewart, Heidi Stigers, Linda Strandenhed, Cristen Stump, Hilde Sture, Rachel Sullivan, Jenni Sutcliffe, Sarah Sutton, Leigh Tawharu, Amanda Taylor, Margaret Taylor, Steve Taylor, Curly Texan, Christian Nedy Theodossy, Anne Thessin, Emily K. Thole, Elizabeth Thomas, Craig Thompson, Hilary Thompson, Natalie Thompson, Stacey Thompson, Ruth Thornton, Financial Times, Liliana Tinoco, Jo Toomey, Suzi Torkington, Jennifer Traeger, Joanna Treggiden, Gwyn Treharne, Lindsay Trevarthen, Sharon Trickett, Jennifer Truax, Aerin Truskey, Adriana Trutzenberg, Kristina Trybek, Yvonne Tsang, Helen Tsepnopoulos, Chloe Tsudama, Gus Tul Laosakul, Kate Turner, Wendy Turner, Lily Turner-Graham, Nikki Turnwald, Mike Tutt, Wendy Tuxworth, Henrietta Ullmann, Eleanor Serafina Unwin, Amandeep Kaur Uppal, Tanya Upton, Haley Urann, Scott Urquhart, Charmi V., Charlotte Vaight, Violeta Valdivia, Leon van Eck, Leah Van Namen, Christopher Vanderkooi, Robbin VanNewkirk, Margaux Vanoosthuyse, Amy Vaughan-Spencer, Jacqueline Ver Meer, Monica Vernay, Armando Villa-Ignacio, Brigette Vinton, Réka Vitz, Stephanie Volk, Shelby Pearl Voorhees, Ellie W, Alma Wahlstrøm Tolsrød, Martha Wailes, Heatherly Walker, Paula Walker, Sir Harold Walker, Jack Wallington, Ben Walsh, Tracy Wang, Debbie Ward, Oliver Ward, Raya Ward, Mark Wardell, Clare Warren, Bas Wassink, Libby Watch, Nicole Watkins Campbell, Angela, Sam, Harriet & Zoe Watling, Edward Webb, Katharina Weinand, Kathryn Weinrauch, Cindy Weisert, Hazel Weller, Matthew Wells, Antje Welzandt, Melanie Wheeler, Corrie Whisner, Deirdre White, Lori White, Alice Whitehead, Liz Whitelegg, Kevin Wigley, Ashley Williams, Bryan Williams, Freya Williams, Sally Williams, Becca Williamson, Heide Wilson, Tony Wilson, Tina Winchester, Mike Witkowski, Melanie Wolf, Alax Wolfden, Amy Womersley, Gillian Wood, Shane Wood, Clare Woodford, Dana Woods, Natasha Woods, Pamela Worley, Lydia Worthington, Callie Wright, Sarah Wright, Willow Wright, Katie Wu, Joseph Wyer, Stephanie Zabinski, Debby Zambo, Scott Zona